Diameter

Diameter

New Generation AAA Protocol – Design, Practice, and Applications

Hannes Tschofenig

Sébastien Decugis

Jean Mahoney

Jouni Korhonen

This edition first published 2019
© 2019 John Wiley & Sons Ltd

The right of Hannes Tschofenig, Sébastien Decugis, Jean Mahoney and Jouni Korhonen to be identified as the authors of this work has been asserted in accordance with law.

Registered Offices
John Wiley & Sons, Inc., 111 River Street, Hoboken, NJ 07030, USA
John Wiley & Sons Ltd, The Atrium, Southern Gate, Chichester, West Sussex, PO19 8SQ, UK

Editorial Office
The Atrium, Southern Gate, Chichester, West Sussex, PO19 8SQ, UK

For details of our global editorial offices, customer services, and more information about Wiley products visit us at www.wiley.com.

Wiley also publishes its books in a variety of electronic formats and by print-on-demand. Some content that appears in standard print versions of this book may not be available in other formats.

Library of Congress Cataloging-in-Publication Data

Names: Tschofenig, Hannes, author.
Title: Diameter : new generation AAA protocol – design, practice, and
    applications / Hannes Tschofenig, Sébastien Decugis, Jean Mahoney, Jouni
    Korhonen.
Description: Hoboken, NJ, USA : John Wiley & Sons, Ltd, [2019] | Includes
    bibliographical references and index. |
Identifiers: LCCN 2019003182 (print) | LCCN 2019009248 (ebook) | ISBN
    9781118875858 (Adobe PDF) | ISBN 9781118875834 (ePub) | ISBN 9781118875902
    (hardcover)
Subjects: LCSH: Diameter (Computer network protocol)
Classification: LCC TK5105.5663 (ebook) | LCC TK5105.5663 .T75 2019 (print) |
    DDC 004.6/2–dc23
LC record available at https://lccn.loc.gov/2019003182

Cover Design: Wiley
Cover Image : © catalby/Getty images

Set in 10/12pt WarnockPro by SPi Global, Chennai, India

Printed in Great Britain by TJ International Ltd, Padstow, Cornwall

10   9   8   7   6   5   4   3   2   1

This book is dedicated to the next AAA.
Hannes, Sébastien, Jean, Jouni

Contents

Disclaimer

This book is based on the authors' personal experiences in the technical field and public standards documents created by the 3rd Generation Partnership Project (3GPP), the Internet Engineering Task Force (IETF), and other standards development organizations. The opinions and views of the authors are solely those of the authors and do not necessarily represent the views of organizations where the authors work. Throughout this book the authors have attempted to make it clear when something is an opinion or a view of the authors. Some of the examples, feature lists, and identified ambiguities may not apply universally to all deployments and products.

The publisher and the authors make no representations or warranties with respect to the accuracy or completeness of the contents of this work and specifically disclaim all warranties, including without limitation warranties of fitness for a particular purpose. No warranty may be created or extended by sales or promotional materials. The advice and strategies contained herein may not be suitable for every situation. This work is sold with the understanding that the publisher is not engaged in rendering legal, accounting, or other professional services. If professional assistance is required, the services of a competent professional person should be sought. Neither the publisher nor the authors shall be liable for damages arising herefrom. The fact that an organization or website is referred to in this work as a citation and/or a potential source of further information does not mean that the authors or the publisher endorses the information that the organization or website may provide or recommendations it may make. Further, readers should be aware that Internet websites listed in this work may have changed or disappeared between when this work was written and when it is read.

About the Authors

This book is a collaborative effort of the following four persons (in alphabetical order):

- **Sébastien Decugis** is the original author and maintainer of the `freeDiameter` implementation. He was involved in IETF activities related to Diameter from 2008 to 2010, and he met Hannes and Jouni working on Diameter at IETF. His work was supported by the National Institute of Information and Communications Technology (NICT) and the WIDE Project, a cross-company and cross-university research project in Japan. Sébastien is not working for those structures anymore, but maintaining and developing the `freeDiameter` implementation on his free time, while NICT and WIDE are kindly maintaining the server resources required by the project.
- **Jouni Korhonen**, PhD, is a Principal R&D Engineer with Nordic Semiconductor. Previously he was with Broadcom and was active in Ethernet-based base station architectures, Ethernet-based fronthaul networks, and time-sensitive networking. Prior to Broadcom Dr. Korhonen was with Nokia Siemens Networks and TeliaSonera where he was heavily involved in IPv6, DNS, mobility, and Diameter-based core network signaling matters in 3GPP and IETF. He has also held multiple leadership positions within IETF and IEEE during his career, including chairing both the Diameter Maintenance and Extensions (DIME) and RADEXT WGs. Dr. Korhonen is still an active contributor with 38 published Requests for Comments (RFCs) to date. His current focus is on cellular Internet of Things and 3GPP system architecture evolution.
- **Jean Mahoney** had wanted to work on an oceanographic research vessel ever since she was a child. Having achieved her life goal right out of college, she discovered that she did not like being in the middle of the ocean. Thus she started on her career in computer networking and technical communication, explaining complex software systems to other developers, system administrators, and users. Jean has more than a decade's worth of experience with IETF specifications and the servers and clients built on top of them. Jean is currently the co-chair of the IETF SIPCORE working group and Gen-ART Secretary.
- **Hannes Tschofenig** co-chaired the IETF DIME working group from March 2006 till March 2010, when he was elected to the Internet Architecture Board (IAB) of the IETF. He is co-author of over 80 RFCs, including several Diameter specifications. His work on Diameter in the IETF was sponsored by Siemens (from 2001 to 2007) and later by Nokia Siemens Networks (from 2007 to 2013), i.e., telecommunication equipment manufacturers selling Diameter-based products. Hannes is now employed by Arm Ltd., where his focus is on improving the security of Internet of Things devices.

Foreword

The Diameter effort started 20 years ago. Its roots were in the limitations of other technologies for user authentication in access networks.

The need for backend authentication servers to assist access networks had grown in the preceding era of modem pools and dial-in access servers. Tools for the simple authentication task existed, but mobile networks in particular needed tools that were capable of growing beyond this task.

The Diameter protocol was born in the IETF. It was designed to be a general-purpose Authentication, Authorization, and Accounting (AAA) protocol for many uses.

Like many other pieces of Internet technology, AAA is not really visible to the end user, but it is crucial for the functioning of the Internet, in particular for the various access networks that provide connectivity for users. When your phone connects to a mobile network, the mobile network's control functions that are needed in the background are built on Diameter.

Mobile networks are the prime area where Diameter is used; simpler other tools also continue their existence and are today equally broadly used, in wireless local area network authentication for instance. And the technology continues to evolve, with most recent designs starting to employ web-based protocols.

The authors, Jean, Jouni, Sébastien, and Hannes, have all worked tirelessly for many years – or even decades – on Diameter, AAA, and other critical Internet technologies. They have written and reviewed specifications, worked on improvements, chaired working groups, built open source implementations, and helped the industry use this technology.

I am happy to see this book on Diameter come out. It focuses on the protocol itself, of course, but also covers open source systems. This is important, as we need both specifications and code to achieve something. It is code that ultimately provides a function, while specifications are needed to ensure that different systems interoperate. The authors approach this book as they have approached their work at the IETF and open source communities, by meticulous attention to detail while keeping the big picture and practical use also up front. Thank you!

Jari Arkko
Senior Expert, Ericsson Research

Preface

Why Did We Write This Book?

"To make money," you will say. That was, however, not our motivation.

We all have been working on Diameter for several years in different roles and therefore we are regularly involved in discussions about Diameter specification questions, or questions about system design and implementation. Often, we go through the same discussions again and again – just with different people.

There was, however, no book to recommend to our co-workers and friends. While we were attending the IETF #86 meeting in Orlando, we sat together outside the conference venue and talked about how to address the common questions we receive. The idea to write a book was born. We reached out to Sébastien, who maintains the free-Diameter implementation, and asked him if he would be willing to help us by creating freeDiameter examples since we prefer hands-on examples in our technical books.

Since two of us had worked with Wiley before on other book projects, we reached out to Wiley again to socialize our idea. To keep it short, you are now holding a Diameter book in your hands.

We hope you enjoy our approach in making you a Diameter expert. We have set up a dedicated website with additional material for this book. If you have questions or feedback, please send us an email at (diameter.book@gmail.com) or visit our webpage at https://diameter-book.info.

What Does This Book Provide?

This book provides the necessary material to understand Diameter, Diameter applications, and the interactions many applications have with the backend infrastructure.

This book provides a coherent picture of Diameter without regurgitating the specifications. It provides information necessary to understand Diameter. To provide you with a hands-on experience and to make your reading experience more interesting, we make use of an open source implementation of the Diameter protocol, called freeDiameter, to

- help you to understand how a Diameter implementation works, and
- illustrate a number of examples using freeDiameter.

It is not our goal to cover everything found in the Diameter specifications. We will, however, provide you with the necessary pointers for further reading.

Who is the Intended Audience?

This book assumes only basic familiarity with how Internet protocols work, such as the concept of IP addresses, the layered protocol stack, and the functions of the layers (particularly the "network layer", the "transport layer", and the "application layer").

We use freeDiameter for examples and to illustrate test setups. A basic understanding of Unix is required in order to set up the freeDiameter environment and to execute the protocol runs. An understanding of TCP/IP will make the examples easier to follow. Readers may skip the examples, but we do recommend engineers use the hands-on experience to gain a deeper understanding of the protocol.

While a technical background or interest in technical matters is a plus, familiarity with the standardization work in the IETF or 3GPP is not required to understand this book.

We believe the following groups will benefit:

- System architects and system designers who have to understand a range of technologies to solve specific use cases. The challenge for those people is to understand the big picture and enough details to glue different protocols together.
- Programmers who need to understand the bigger picture of the Diameter protocol.
- Standardization experts who are new to Diameter or need to define new Diameter extensions.
- Students and researchers who are interested in technology that is deployed by many network operators. Typically, Diameter is not widely known since it is not an end-user-facing technology.
- Technical marketing people who want to gain a better understanding of the technology they are dealing with.

Book Structure

Chapter 1 discusses the motivation for using Diameter, briefly talks about the predecessor to Diameter, RADIUS, and introduces the open source Diameter implementation, freeDiameter.

Chapter 2 describes Diameter via its building blocks. These building blocks are then used to illustrate the basic peer-to-peer communication between neighboring Diameter nodes in Chapter 3.

Chapter 4 extends Diameter communication from two nodes to an arbitrary number of nodes.

Following the chapters covering communication between Diameter nodes, Chapter 5 introduces security functionality. Diameter security is today mainly implemented and deployed at the level of peer-to-peer communication.

Chapter 6 describes selected Diameter applications in more detail. We have chosen applications that are deployed today and illustrate the flexibility and capabilities of Diameter well.

Chapter 7 is an advanced chapter that teaches you how to develop your own Diameter extensions, for example by defining new attribute–value pairs (AVPs), new commands, or even completely new Diameter applications.

We have added `freeDiameter` examples throughout the book as far as applicable. Not every standardized functionality is already available in `freeDiameter`.

Acknowledgements

This book effort took much longer than we had expected. A big thanks to Wiley for their continued support and patience. We (Jean, Sébastien, Jouni, and Hannes) had specific ideas in mind of what type of book we wanted to write. We are happy that we managed to implement our ideas in this book without taking shortcuts and came this far in our journey.

Needless to say, this book project would not have been possible without the efforts put forth by those working in the IETF DIME working group. We would also like to thank our peers working in other organizations on Diameter extensions, specifically in 3GPP. A tremendous amount of work has gone into Diameter's standardization and widespread deployment. It does not just happen by accident or luck.

Writing more than 200 pages was not easy, and we would like to thank our families for their patience. Without their support it would not have been possible to complete this project. Thank you, Verena, Elena, Robert, and Hanna. Jouni also sends special thanks to Dana Street Roasting Company for its excellent caffeine-rich products that helped him to stay focused as the writing of this book took place during hours when normal people sleep.

Finally, we would also like to thank our employers. They have enabled us to participate in various standards developing organizations for many years. Not only have we been able to work on exciting technical topics, and to travel around the world to participate in many face-to-face standardization meetings and interoperability events, but we also met many great people.

Hannes, Sébastien, Jean, Jouni

List of Abbreviations

ABNF	Augmented Backus–Naur Form
AE	authorizing entity
AESE	Architecture Enhancements for Service Capability Exposure
AoC	advice of charge
AppE	application endpoint
CA	certification authority
CCF	Command Code Format
CEA	Capability Exchange Answer
CER	Capability Exchange Request
CIoT	Cellular Internet of Things
COPS	Common Open Policy Service Protocol
CRL	certificate revocation list
DCN	dedicated core network
DDDS	Dynamic Delegation Discovery Service
DEA	diameter edge agent
DIME	Diameter Maintenance and Extensions
DNS	Domain Name System
DOIC	Diameter Overload Information Conveyance
DPA	Disconnect Peer Answer
DPR	Disconnect Peer Request
DRA	diameter routing agent
DRMP	Diameter Routing Message Priority
DSL	digital subscriber line
DTLS	Datagram Transport Layer Security
DWA	Device Watchdog Answer
DWR	Device Watchdog Request
E-UTRAN	Evolved Universal Terrestrial Radio Access Network
EAP	Extensible Authentication Protocol
ECC	elliptic curve cryptography
EPC	Evolved Packet Core
EPS	Evolved Packet System
FQDN	fully qualified domain name
GSMA	GSM Association
HSS	Home Subscriber Server
HTTP	Hypertext Transfer Protocol

IANA	Internet Assigned Numbers Authority
IESG	Internet Engineering Steering Group
IKEv2	Internet Key Exchange version 2
IMEI-SV	International Mobile Station Equipment Identity and Software Version number
LTE	Long-Term Evolution
MME	Mobility Management Entity
MPTCP	Multipath TCP
NAI	Network Access Identifier
NAPTR	Naming Authority Pointer
NAS	network access server
NE	network element
NTP	Network Time Protocol
OCS	overload control state
OCSP	Online Certificate Status Protocol
OLR	overload report
OVF	Open Virtualization Format
PDN	packet data network
PDP	Packet Data Protocol
PKI	public key infrastructure
QoS	quality of service
RADIUS	Remote Authentication Dial In User Service
RAT	radio access technology
RR DNS	Resource Record
RRE	resource-requesting entity
S-NAPTR	Straightforward-NAPTR
SCEF	Service Capability Exposure Function
SCM	source code management
SCTP	Stream Control Transmission Protocol
SDO	standards development organizations
SGSN	Serving GPRS Support Node
SGW	signaling gateway
SIP	Session Initiation Protocol
SLPv2	Service Location Protocol version 2
SNMP	Simple Network Management Protocol
SRV	Service Location
SS7	Signaling System 7
TCP	Transmission Control Protocol
TLS	Transport Layer Security
UDP	User Datagram Protocol
UE	user equipment
URI	Uniform Resource Identifier
V2X	vehicle-to-everything communication
VM	virtual machine
VoIP	voice-over IP
VPN	virtual private network
WLAN	wireless local area network

1

Introduction

1.1   What is AAA?

AAA stands for *Authentication, Authorization*, and *Accounting*.

Authentication is the verification that a user who is requesting services is a valid user of the network services requested. The user must present an identity, like a user name or phone number, and credentials, like a password, a digital certificate, or one-time passphrase, to the verifier in order to be authenticated.

Authorization is the determination of whether requested services can be granted to a user who has presented an identity and credentials based on their authentication, service request, and system state. Authorization state may change over the course of a user's session due to consumption limits or time of day.

Accounting is the tracking of the user's consumption of resources for billing, auditing, and/or system planning. Typical accounting data collected includes the identity of the user, the service delivered, and when the service started and stopped.

Consider a voice-over IP (VoIP) service provider that offers telephony services to a large number of end users. End users can connect to the service with software for VoIP clients that runs on a smart phone, tablet or desktop PC, or they may use a purpose-built hardware phone.

When the user's device contacts the VoIP network, the VoIP service provider will *authenticate* the user accessing their network. That is, the provider wants to determine that the user, or her device, is who they say they are. The authentication mechanisms and credentials vary by deployment. For example, some deployments may use human-memorizable username and password combinations, while others may use a public key infrastructure with certificates stored on smart cards.

Once the VoIP service provider has successfully authenticated the user, the provider will then *authorize* them to use the services by verifying the conditions and privileges of the user's account and the status of the user's credits for the requested action, such as making a phone call.

If the user successfully passes the authorization procedure, the user's resource consumption will be *accounted*. Accounting resource consumption is useful for a number of reasons, including capacity planning, understanding user behavior to improve service experience, charging for service use, and measuring policy compliance. The kinds of data collected as part of the accounting process depend on the application

Diameter: New Generation AAA Protocol – Design, Practice, and Applications, First Edition.
Hannes Tschofenig, Sébastien Decugis, Jean Mahoney and Jouni Korhonen.
© 2019 John Wiley & Sons Ltd. Published 2019 by John Wiley & Sons Ltd.

context and the needs of the service provider, and the data may need to be collected from various places in the network. For example, one VoIP service provider may collect data about transmitted voice packets. Another provider may be satisfied with collecting data about the call setup procedures only.

Typically VoIP deployments use Session Initiation Protocol (SIP) for call setup. In small VoIP deployments that use SIP, the AAA operations happen within the SIP proxy, which is a network element that helps to route SIP requests to their final destinations. As a SIP network grows larger, the VoIP service provider may deploy a dedicated and centralized AAA server to manage subscribers' information and their authorization properties on behalf of multiple proxies. When a service request arrives at a SIP proxy, the proxy will send AAA-related requests to the AAA server.

The SIP proxy in this distributed network is a kind of network access server. Network access server (NAS) is a generic term for the end user's entry point to a network. A NAS provides services on a per-user basis, based on authentication, and ensures the service provided is accounted for. A NAS contacts a separate AAA server to verify the user's credentials and then sends accounting data to the AAA server. A NAS, then, is an AAA client.

When the AAA functionality is outsourced from a NAS to the AAA server, there needs to be a protocol defined between the AAA client within the NAS and AAA server. Since the developers who created the NAS are likely different than the developers who created the AAA server, it is helpful to not only define a communication protocol, but also to agree on an open standard rather than to use a proprietary interface. In fact, various AAA protocol standards have been defined, with standards work starting with the early Internet dial-up services and progressing to cover connections to today's modern wireless networks.

1.2   Open Standards and the IETF

The standards organization that works to improve the interoperability of the Internet is the Internet Engineering Task Force (IETF), an international community of network designers, operators, vendors, and researchers that develop open, voluntary Internet standards. Examples of such standards include Internet transport (TCP/IP, UDP), email (SMTP), network management (SNMP), web (HTTP), voice over IP (SIP), and also AAA (RADIUS, Diameter). The IETF does not have formal membership requirements and is open to anyone interested in improving the Internet. The newcomer's guide to the IETF is known as *The Tao of the IETF* [1] and can be found online.

Standards work in the IETF is done in working groups, which discuss protocol solutions on mailing lists and in person at IETF meetings, and capture these solutions in documents known as Internet drafts. Working groups are self-organized by topic and are grouped into broad focus areas. Work on AAA protocols has taken place in multiple working groups.

The gauge of a protocol in the IETF is "rough consensus and running code". When the working group has arrived at rough consensus, the Internet draft enters a review period known as a Last Call, in which the larger IETF community can provide input. Internet drafts are then reviewed by the Internet Engineering Steering Group (IESG). When the IESG approves an Internet draft, the draft moves on to become a Request for

Comments (RFC), which, despite its categorization, is now at a level of stability that it can be implemented with confidence.

The details of IETF Internet protocols, such as port numbers, application identifiers, and header field names, are stored with the Internet Assigned Numbers Authority (IANA), which is responsible for the global coordination of Internet protocol resources.

1.3 What is Diameter?

Diameter is an open standard AAA protocol defined by the IETF. Diameter's features fulfill multiple requirements of network operators. The definition of the Diameter protocol is given in the Diameter base specification, RFC 6733 [2].

Various AAA protocols, such as the Common Open Policy Service Protocol (COPS) [3] and Remote Authentication Dial In User Service (RADIUS) [4], had been developed before work on the Diameter protocol started. Experience with these protocols provided the IETF community with requirements for a next-generation AAA protocol. These requirements are documented in RFC 2989, *Criteria for Evaluating AAA Protocols for Network Access* [5]. The design of Diameter incorporated the lessons learned from these various AAA protocols.[1]

As work continued on Diameter, the AAA working group of the IETF [6] evaluated the available AAA protocols against the requirements given in RFC 2989. Those requirements are:

Scalability: The AAA protocol has to be able to support millions of end users and tens of thousands of devices, AAA servers, Network Access Servers, and brokers.

Failover: Failover support aims to provide uninterrupted AAA service in the case of a failure. Failover requires the detection of a failed node and the re-routing of outstanding messages to an alternative node. Failover support may lead to the retransmission of messages, and those duplicate messages should be handled appropriately by the protocol.

Security: AAA protocols carry sensitive data, including long-term authentication credentials (such as passwords), session keys, service usage information as part of accounting records, and possibly the end user's location. From a data protection point of view, this personal data requires special care. Network operators also want to avoid malicious parties injecting false information into the system. For this purpose, providing a common, widely used security mechanism is desirable.

Reliable transport: When accounting records were transmitted over an unreliable transport protocol, as done in earlier protocols (such as RADIUS), packet loss translated to loss of money. Consequently, implementations added their own reliability mechanisms, leading to differences among vendors. Diameter incorporates the reliable transports the Transmission Control Protocol (TCP) and the Stream Control Transmission Protocol (SCTP) to ensure uniform behavior among implementations by different vendors.

Agent support: Earlier AAA protocols did not offer nodes that could route and redirect AAA messages, which can future-proof a AAA network deployment.

---

1 A joke among IETF participants is to point out that diameter is twice the radius. Hence, Diameter is meant to be the next generation AAA protocol after RADIUS.

Server-initiated messaging: In earlier AAA protocols, only the AAA client could initiate the message exchange. The AAA server's ability to initiate messages was added later as an optional feature, and therefore support for it could not be assumed. Server-initiated messaging is used, for example, when authorization characteristics change and re-authorization by the user or the end device is required. With long-running sessions, this initiation has to be triggered by the AAA server towards the AAA client.

Transition support: Since Diameter would be introduced into existing AAA deployments, it needed to provide a transition story to lower the deployment effort for network operators.

Ability to carry service-specific attributes: The AAA protocol needs to be extensible and provide the ability to define new attributes required by new services.

The AAA working group published their results in RFC 3127 [7], *Authentication, Authorization, and Accounting: Protocol Evaluation*, expressing a preference for Diameter since it met most of the requirements specified in RFC 2989 and needed only minor engineering to bring it into complete compliance. Since the Diameter specification was still under development, the working group could address the requirement gaps.

1.3.1 Diameter versus RADIUS

A book about Diameter cannot be silent about its predecessor, RADIUS. RADIUS was originally standardized in January 1997 by the IETF with RFC 2058 [8], which was replaced by RFC 2138 [9] a few months later, and was made obsolete in June 2000 by RFC 2865 [4].

Diameter was able to address deficiencies found in the RADIUS protocol, namely:

- No reliable message delivery. RADIUS used the User Datagram Protocol (UDP), an unreliable transport protocol, to communicate messages from a RADIUS client to a RADIUS server.
- RADIUS had a monolithic design whereby RADIUS attributes used by different applications were put into one bucket for transport with RADIUS. Unlike Diameter, RADIUS did not separate the message delivery from the application's semantics. Interoperability issues and lack of extensibility were common problems in deployments.
- The Datagram Transport Layer Security (DTLS) protocol did not exist at that time, therefore RADIUS had to rely on either IPsec [10] or no security protection at all.
- Only client-initiated messaging. RADIUS initially did not provide a mechanism for letting the server initiate messages.
- RADIUS only supported a basic set of data types, which made it difficult for application designers to define their own RADIUS attributes.

This was, however, not the end of the story since, paralleling the Diameter work within the IETF AAA working group and later continued in the RADIUS [11] and RADEXT [12] working groups, the RADIUS protocol experienced a number of improvements, many of which were inspired by work on the Diameter protocol:

- Reliable message delivery. RFC 6613 [13] added support for the TCP to RADIUS.

- Improved security. RFC 7360 [14] added the ability to use DTLS with UDP-based RADIUS messages. RFC 6614 [15] added support for TLS for TCP-based RADIUS messaging.
- Server-initiated messages. RFC 3576 [16] added support for server-initiated messages as part of the dynamic authorization extensions.
- Extended attribute type space. The demand for attributes had been close to exhausting RADIUS's 8-bit attribute type space. RFC 6929 [17] extended the type space and added a mechanism for complex attributes.
- Design guide. To combat interoperability problems caused by protocol design activities in various organizations, RADIUS design guidelines were published with RFC 6158 [18].

At the time of this writing, development of the RADIUS protocol is still ongoing in the IETF radext working group. However, not only does the IETF develop extensions for RADIUS, but other organizations do also. Hence, the best way to gain an overview of the available extensions is to look at the IANA registry for RADIUS [19].

Today, many of the features of Diameter are also available within RADIUS. It is therefore fair to ask which communities are driving the development of each protocol. It turns out that many small- and medium-size enterprises use RADIUS, including many WLAN hotspot deployments, universities, and digital subscriber line (DSL) and cable operators. On the other hand, large Internet service providers, and particularly mobile operators, use Diameter in their network architectures. The market is therefore is nicely divided, and does not lead to rivalry in the standardization environment.

1.3.2   Diameter Improvements

It is important to note that the Diameter base specification (RFC 6733 [2]) is a revision of the original Diameter protocol, specified in RFC 3588 [20], and is the output of the IETF DIME working group [21], which incorporated feedback of protocol implementers from interoperability testing events and discussions on working group mailing lists. RFC 6733 obsoletes RFC 3588.

The main differences between RFC 3588 and RFC 6733 are the following:

Security: RFC 6733 specifies Transport Layer Security (TLS) (when used with TCP) and DTLS (when used with SCTP) as the primary ways to secure Diameter messages. It also specifies the use of a well-known port, which is similar to how TLS is used with other application-layer protocols such as HTTPS. The end-to-end security framework described in RFC 3588 is deprecated since the actual technical solution has not yet been standardized. More discussion about Diameter security can be found in Chapter 5.

Diameter Node Discovery: A Diameter node discovers to which node it needs to talk via either manual configuration or a dynamic discovery procedure. RFC 6733 simplifies the dynamic discovery procedure since it was observed that many vendors had implemented only the DNS-based mechanism.

Extensibility: The story for extending Diameter presented in RFC 3588 was unclear and led to incompatible extensions. RFC 6733 clarified Diameter extensibility, and Chapter 7 is dedicated to the topic of Diameter extensibility to provide help to those who want to develop their own extensions.

Clarifications: RFC 6733 is full of helpful clarifications for readers and implementers. The clarifications are the results of many discussions within the working group to reconstruct the original intentions and to match them with implementations in the field.

More details about these differences can be found in Section 1.1.3 of RFC 6733.

Given these changes, we recommend that you look at RFC 6733 even though older implementations focus on RFC 3588. It is important to understand that many implementations will need time to meet the additional requirements outlined in RFC 6733. In particular, the security changes will lead to changes in implementation code. It is hoped that, by the time you read this book, many, if not most, vendors will have conducted interoperability tests and therefore have taken the various clarifications into account.

1.4   What is `freeDiameter`?

`freeDiameter` is an open source implementation of the Diameter protocol. Development on `freeDiameter` was started in 2008 as an academic project with the goals of evaluating and promoting the Diameter protocol as specified by RFC 3588. `freeDiameter` has evolved to follow the revisions of the Diameter protocol in RFC 6733, part of which were introduced as a result of the evaluation started with `freeDiameter`.

`freeDiameter` has been used in commercial Diameter deployments, and it can be used as a reference implementation that anyone developing a commercial Diameter stack can use for interoperability testing. It is also a platform made freely available to researchers and students for prototyping, and for evaluating their ideas for new services built upon Diameter. For these reasons, `freeDiameter` was written in the C language and has been engineered to be as flexible and extensible as possible, with a small system footprint and good performance.

We will use `freeDiameter` throughout this book to illustrate various concepts of the Diameter protocol. By following the hands-on examples in this book, `freeDiameter` will give you a better understanding of Diameter as you configure it to exchange Diameter messages between different nodes. Instructions on setting up `freeDiameter` can be found in Appendix A.

References

1 IETF. The Tao of IETF: A Novice's Guide to the Internet Engineering Task Force, Nov. 2012. http://ietf.org/tao.html.

2 V. Fajardo, J. Arkko, J. Loughney, and G. Zorn. Diameter Base Protocol. RFC 6733, Internet Engineering Task Force, Oct. 2012.

3 D. Durham, J. Boyle, R. Cohen, S. Herzog, R. Rajan, and A. Sastry. The COPS (Common Open Policy Service) Protocol. RFC 2748, Internet Engineering Task Force, Jan. 2000.

4 C. Rigney, S. Willens, A. Rubens, and W. Simpson. Remote Authentication Dial In User Service (RADIUS). RFC 2865, Internet Engineering Task Force, June 2000.

5 B. Aboba, P. Calhoun, S. Glass, T. Hiller, P. McCann, H. Shiino, P. Walsh, G. Zorn, G. Dommety, C. Perkins, B. Patil, D. Mitton, S. Manning, M. Beadles, X. Chen, S. Sivalingham, A. Hameed, M. Munson, S. Jacobs, B. Lim, B. Hirschman, R. Hsu, H. Koo, M. Lipford, E. Campbell, Y. Xu, S. Baba, and E. Jaques. Criteria for Evaluating AAA Protocols for Network Access. RFC 2989, Internet Engineering Task Force, Nov. 2000.

6 IETF. Authentication, Authorization and Accounting (AAA) (Concluded) Working Group, Mar. 2014. http://datatracker.ietf.org/wg/aaa/charter/.

7 D. Mitton, M. St.Johns, S. Barkley, D. Nelson, B. Patil, M. Stevens, and B. Wolff. Authentication, Authorization, and Accounting: Protocol Evaluation. RFC 3127, Internet Engineering Task Force, June 2001.

8 C. Rigney, A. Rubens, W. Simpson, and S. Willens. Remote Authentication Dial In User Service (RADIUS). RFC 2058, Internet Engineering Task Force, Jan. 1997.

9 C. Rigney, A. Rubens, W. Simpson, and S. Willens. Remote Authentication Dial In User Service (RADIUS). RFC 2138, Internet Engineering Task Force, Apr. 1997.

10 S. Kent and K. Seo. Security Architecture for the Internet Protocol. RFC 4301, Internet Engineering Task Force, Dec. 2005.

11 IETF. Remote Authentication Dial-In User Service (RADIUS) (Concluded) Working Group, Mar. 2000. http://datatracker.ietf.org/wg/radius/charter/.

12 IETF. RADIUS EXTensions (RADEXT) Working Group, Mar. 2014. http://datatracker.ietf.org/wg/radext/charter/.

13 A. DeKok. RADIUS over TCP. RFC 6613, Internet Engineering Task Force, May 2012.

14 A. DeKok. Datagram Transport Layer Security (DTLS) as a Transport Layer for RADIUS. RFC 7360, Internet Engineering Task Force, Sept. 2014.

15 S. Winter, M. McCauley, S. Venaas, and K. Wierenga. Transport Layer Security (TLS) Encryption for RADIUS. RFC 6614, Internet Engineering Task Force, May 2012.

16 M. Chiba, G. Dommety, M. Eklund, D. Mitton, and B. Aboba. Dynamic Authorization Extensions to Remote Authentication Dial In User Service (RADIUS). RFC 3576, Internet Engineering Task Force, July 2003.

17 A. DeKok and A. Lior. Remote Authentication Dial In User Service (RADIUS) Protocol Extensions. RFC 6929, Internet Engineering Task Force, Apr. 2013.

18 A. DeKok and G. Weber. RADIUS Design Guidelines. RFC 6158, Internet Engineering Task Force, Mar. 2011.

19 IANA. Radius Types, Jan. 2016. https://www.iana.org/assignments/radius-types/radius-types.xhtml.

20 P. Calhoun, J. Loughney, E. Guttman, G. Zorn, and J. Arkko. Diameter Base Protocol. RFC 3588, Internet Engineering Task Force, Sept. 2003.

21 IETF. Diameter Maintenance and Extensions (DIME), Mar. 2014. http://datatracker.ietf.org/wg/dime/charter/.

2

Fundamental Diameter Concepts and Building Blocks

2.1 Introduction

This chapter covers the basic concepts of Diameter, describing the types of nodes that participate in Diameter sessions and the format of Diameter messages. It also provides an overview of Diameter sessions and error handling.

2.2 Diameter Nodes

As mentioned in Chapter 1, Diameter nodes are auxiliary participants in larger networks that offer various services to end users. A Diameter node is a software application that implements the Diameter protocol to provide authentication, authorization, and accounting services to entities or devices requesting or using network services. Two Diameter nodes that share a direct transport connection with each other are called Diameter peers. A Diameter node can act as a client, server or agent:

Client
A Diameter client creates Diameter messages to request authentication, authorization, and accounting services for network users. The behavior of a Diameter client depends on the application it is supporting. A Diameter client is sometimes known as a network access server, which can be confusing. However, from a user perspective, the NAS is a server since a user interacts with it to gain access to the network, but within the Diameter framework it is an application client.

Server
A Diameter server receives and handles authentication, authorization, and accounting requests for a particular realm or administrative domain. A Diameter server supports specific Diameter applications, in addition to the base protocol. Like Diameter clients, the behavior of the Diameter server depends on the Diameter applications it supports.

Agent
A Diameter agent is an intermediary that assists with the communication between clients and servers. Agents may simply forward messages or they may make policy decisions about the messages they handle. Other agents provide protocol translation services. An agent may combine one or more agent roles.

Diameter: New Generation AAA Protocol – Design, Practice, and Applications, First Edition.
Hannes Tschofenig, Sébastien Decugis, Jean Mahoney and Jouni Korhonen.
© 2019 John Wiley & Sons Ltd. Published 2019 by John Wiley & Sons Ltd.

2.3   Diameter Protocol Structure

The Diameter protocol design follows a two-layer structure, as shown with two Diameter nodes (Figure 2.1). The lower layer is the peer-to-peer messaging layer where Diameter peers connect and discover each other's capabilities. The upper layer is the Diameter application layer, where Diameter nodes exchange AAA data tailored to specific application needs. In this figure the roles of the nodes, whether they are a client, server, or an agent, are abstracted away.

Figure 2.2 illustrates the Diameter protocol's relationship to an application server that triggers the AAA interaction. The user requests service from an application server. The application server checks the user's credentials via Diameter. The figure is simplified with respect to Diameter nodes. Only the Diameter client and a Diameter server are shown; any Diameter agents in the path have been omitted. The Diameter client can be either separate or integrated with an application. Here we show that the application server communicates with the client via an application programming interface.

2.4   Diameter Applications

Diameter is a flexible, extensible protocol. The base protocol specifies only a basic accounting application (Table 2.1), but provides building blocks for creating new

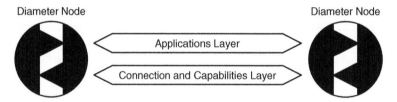

Figure 2.1 The layered design of Diameter.

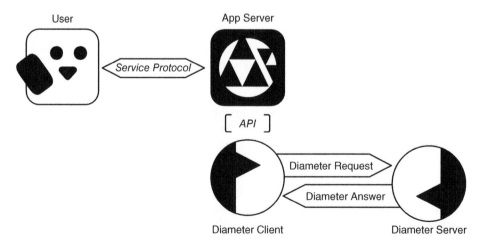

Figure 2.2 Diameter providing AAA support for a generic service protocol.

Table 2.1 Applications offered by the Diameter base protocol.

Application	Application-Id
Common message	0
Base accounting	3
Relay	4294967295

applications that can address the particulars of authentication, authorization, and/or accounting for various networks. Other Diameter applications are specified in separate documents created by the IETF and other organizations, and some of these applications are covered in depth in Chapter 6. A list of over 150 Diameter applications with links to their specifications can be found at IANA [1].

Within Diameter messages, applications are noted with application identifiers. Diameter nodes inform each other of the applications that they support. For instance, a client advertises that it supports the Diameter quality of service (QoS) application (Application-Id 9 [2]) in its request, but if the server does not support the QoS application, it would leave it out of its answer. Since it is unhelpful for a Diameter application request to reach a server that cannot handle that particular application, the application information within Diameter messages is also used for routing decisions.

2.5 Connections

A Diameter connection is a transport-level connection between two peers that is used to send and receive messages. This section provides a brief overview. Connections are covered in greater detail in Chapter 3.

2.5.1 Transport Layer

A Diameter connection with a peer starts with a Diameter node setting up a transport connection with that peer using either TCP [3] or SCTP [4]. TCP and SCTP both provide acknowledged, error-free, non-duplicated transfer of user data. Both provide reliable data transfer and strict order-of-transmission delivery of data. SCTP adds a few features to TCP that allow better failover to improve resiliency and increased denial-of-service resistance.

When TCP is used, the connections should be secured with TLS [5]. When SCTP is used, the connections should be secured with DTLS [6]. While a Diameter client can use either TCP or SCTP, agents and servers should support both transports.

TLS/TCP and DTLS/SCTP are the primary methods of securing Diameter. IPsec, specified in RFC 4301 [7], is an alternative for securing connections between peers if only TCP or SCTP is used. For more information about security, see Chapter 5.

A Diameter node may initiate connections from a source port different from the one that it listens on, but it must be prepared to receive connections on the default ports. The following are the default ports for Diameter:

- 3868 for TCP and SCTP
- 5868 for TLS and DTLS.

Note that RFC 6733 defines port number 5868 for secure transport in Section 11.4, IANA Considerations, and that port number was added to the IANA port numbers registry. However, the text in Sections 2.1 and 4.3.1 of RFC 6733 gives an incorrect port number. The correct port for secure transports is 5868.

2.5.2 Peer-to-Peer Messaging Layer

Once the transport connection is established, a Diameter node sends to the peer a Capabilities-Exchange-Request (CER) that contains information about the node's identity and the Diameter applications and security that the node supports. The peer returns a Capabilities-Exchange-Answer (CEA), with its own identity and supported applications.

Connections are maintained with Device-Watchdog-Request/Device-Watchdog-Answer exchanges, which are used to detect connection failure. A node sends a Device-Watchdog-Request to its peer when there has been no traffic over the connection for some period of time. The peer responds with a Device-Watchdog-Answer. If the requesting node does not receive an answer, it may either send queued messages to an alternate peer and alert the peer that the messages may be potentially retransmitted, or it may tell downstream[1] peers that it was unable to deliver the messages. The node will also periodically attempt to reconnect with the peer by sending CERs.

A connection is shut down with a Disconnect-Peer-Request/Disconnect-Peer-Answer exchange. The Disconnect-Peer-Request allows a node to inform its peer that it plans to tear down the transport connection. The node can indicate one of three reasons for the disconnection: rebooting, busy, or the connection is no longer necessary. The node waits until it receives the Disconnect-Peer-Answer before tearing down the connection since the peer can warn the node in its answer if there are messages in flight.

2.5.3 Setting up a Connection between `freeDiameter` Peers

If you have not already done so, turn to Appendix A to learn how to set up `freeDiameter`. The appendix contains a simple example of setting up a connection between two peers.

2.6 Diameter Message Overview

A Diameter message, also known as a Diameter command, is either a request or an answer to a request, and consists of a header followed by a payload of various attribute–value pairs (AVPs). For example, a CER and a CEA are sent after two Diameter nodes establish a transport connection. In the CER, the node sends information to

---

1 The terms "upstream" and "downstream" are used occasionally within specifications to identify the direction of a Diameter application message. The term "upstream" identifies the direction from the Diameter client toward the server that will handle the request. The term "downstream" identifies the direction of a Diameter message from the answering server toward the Diameter client.

Table 2.2 Commands defined by the Diameter base protocol.

Command name	Abbreviation	Code
Abort-Session-Request	ASR	274
Abort-Session-Answer	ASA	274
Accounting-Request	ACR	271
Accounting-Answer	ACA	271
Capabilities-Exchange-Request	CER	257
Capabilities-Exchange-Answer	CEA	257
Device-Watchdog-Request	DWR	280
Device-Watchdog-Answer	DWA	280
Disconnect-Peer-Request	DPR	282
Disconnect-Peer-Answer	DPA	282
Re-Auth-Request	RAR	258
Re-Auth-Answer	RAA	258
Session-Termination-Request	STR	275
Session-Termination-Answer	STA	275

its peer about its identity and which Diameter applications and security that it supports. The peer returns to the client a CEA, with its own identity and supported applications.

Table 2.2 lists the commands defined by the Diameter base protocol. Note that a command request and and its corresponding answer are tied together with, and identified by, a unique number known as a Command Code. The Command Code for the CER and CEA pair is 257.

2.6.1 The Command Code Format

When Diameter commands are discussed in specifications, they are presented in a human-readable format known as a Command Code Format (CCF). CCF is based on the Augmented Backus-Naur Form (ABNF) metalanguage [8].

The example of CCF given in Figure 2.3 is the definition of the CER message format.

The first line provides the command name or abbreviation. In this example, the abbreviation CER is used. Two colons and an equal sign (: : =) separate the command name from the Diameter Header, which contains the Command Code (257, Table 2.2).

The header also contains the Command Flags field. Here, REQ indicates that the message is a request, that is, the r-bit is set. PXY or ERR are given if the message is proxiable (p-bit is set) or if the message contains a protocol error (e-bit is set), respectively. Table 2.3 provides more information about Command Flags.

Below the header line of the CCF is a list of AVPs that the message may contain. AVPs shown in curly braces ({ }) are mandatory. AVPs shown in square brackets ([ ]) are optional. AVPs shown in angle brackets (<>) have a fixed position in the CCF thus effectively making them also mandatory. Fixed position AVPs have to be positioned in the CCF in such a way that their position will not change. Asterisks indicate that an AVP can appear more than once. A number to the left of the asterisk indicates the minimum number of times an AVP may appear; a number to the right indicates a maximum. If

```
<CER> ::= < Diameter Header: 257, REQ >
              { Origin-Host }
              { Origin-Realm }
        1* { Host-IP-Address }
              { Vendor-Id }
              { Product-Name }
              [ Origin-State-Id ]
           * [ Supported-Vendor-Id ]
           * [ Auth-Application-Id ]
           * [ Inband-Security-Id ]
           * [ Acct-Application-Id ]
           * [ Vendor-Specific-Application-Id ]
              [ Firmware-Revision ]
           * [ AVP ]
```

Figure 2.3 An example of the Command Code Format.

Table 2.3 Meanings of Command Flag fields.

Bit	ID	Meaning when Set	Meaning when Cleared
0	R	Message is a **R**equest	Message is an answer
1	P	Message may be **P**roxied, relayed, or redirected	Message must be processed locally
2	E	Request caused a protocol **E**rror	Cleared in initial requests and in answers that report transient errors or permanent failures
3	T	A resent, unacknowledged request. Indicates possible duplication due to a link failure.	Cleared in initial requests or answers
4	r	Should be ignored	reserved for future use
5	r	Should be ignored	reserved for future use
6	r	Should be ignored	reserved for future use
7	r	Should be ignored	reserved for future use

there is no number given before the asterisk, it can be assumed that the value is 0 if the AVP is optional or 1 if the AVP is required. If there is no number given after the asterisk, it is assumed that the maximum is infinity.

Note that the Vendor-Specific-Application-Id AVP is of type Grouped (Table 2.5). A Grouped AVP is a concatenation of zero or more AVPs. The CCF for this grouped AVP can be seen in Figure 2.4.

```
<Vendor-Specific-Application-Id> ::= < AVP Header: 260 >
                                        { Vendor-Id }
                                        [ Auth-Application-Id ]
                                        [ Acct-Application-Id ]
```

Figure 2.4 An example of a CCF for a Grouped AVP.

The last line of the example, * [AVP], indicates that the message is extensible and that new AVPs may be added by specification at a later time. Extensibility is covered in detail in Chapter 7.

2.6.2 Message Structure

When Diameter messages are transmitted between nodes, they are transmitted in network byte order, that is, with the most significant byte sent first. Figure 2.5 shows how the message is structured for transmission.

The Diameter message contains the following:

Version
The first field in a Diameter message is Version. Since there is only the first version of Diameter as of this writing, this field is always set to 1.

Message Length
The Message Length field indicates the number of octets of the Diameter message, including the header fields and the padded AVPs.

Command Flags
Command Flags indicate the type of Diameter message and how the message should be handled (Table 2.3). The "R" bit, indicating whether the message is a request, and the "P" bit, indicating whether a message can be proxied, are straightforward. The "E" or Error bit is nuanced. Setting the Error bit can indicate a certain type of error, but the bit can be clear with other sorts of errors, and can be set when an agent redirects a request, which is not necessarily an error. More details on error handling are given in Section 2.8. Chapter 4 covers the details of redirecting requests.

Command Code
The Command Code field contains the command associated with the message. Table 2.2 lists Command Codes that are defined in the Diameter base protocol. IANA maintains the registry of Diameter Command Codes [1].

Application-Id
The Application-Id message field identifies the application to which the message is applicable. A list of applications defined by the base protocol and their Application-Ids is given in Table 2.1. IANA maintains the registry of Diameter applications and their Application-Ids [1].

Hop-by-Hop Identifier
The Hop-by-Hop Identifier is a field used to match answers to requests. The sending node includes a unique Hop-by-Hop Identifier in its request. The peer sends an answer using the same Hop-by-Hop Identifier value found in the corresponding

Figure 2.5 Diameter message format.

Version	Message Length		
Cmd Flags	Command Code		
Application-Id			
Hop-by-Hop Identifier			
End-to-End Identifier			
AVP(s)...			

0   8   16   24   32

request. If the message is relayed or proxied, the agent will save the Hop-by-Hop Identifier, and replace the value with its own. When the agent receives an answer, it replaces the Hop-by-Hop Identifier with the downstream peer's earlier saved identifier. The Hop-by-Hop Identifier is usually a monotonically increasing number, starting from a randomly generated number. If a node receives an answer message with an unknown Hop-by-Hop Identifier, it discards the message.

End-to-End Identifier

The End-to-End Identifier is a field that is used to detect duplicate messages. The node inserts a unique End-to-End Identifier in its request, which is preserved even if the message is relayed or proxied. The originator of an answer message uses the same End-to-End Identifier field found in the request. The combination of the `Origin-Host` AVP and this field is used to detect duplicate messages.

AVPs

The AVPs are the payload of the Diameter message. Which AVPs are included in a message will vary with each Diameter command. The format of an AVP depends on its definition. AVP formats are covered in Section 2.6.3.

2.6.3 Attribute–Value Pairs

AVPs contain the payload of the Diameter message. Tables 2.4 and 2.5 list the AVPs that are defined by the base protocol. The AVP code identifies the AVP, the data type specifies the size and type of data the AVP carries, and the AVP flag of "M" means that the receiver needs to be able to understand the AVP in order to process the Diameter message.

2.6.3.1 Format

The format of the AVP header is shown in Figure 2.6. The fields in the AVP header are sent in network byte order.

AVP Code field

The AVP Code plus the optional Vendor-Id field identify the attribute. AVP numbers 1 through 255 are reserved for reuse of RADIUS attributes. AVP numbers 256 and above are used for Diameter. IANA maintains the list for AVP Codes that have been defined by the IETF [1].

AVP Flags

The AVP Flags field informs the receiver how each attribute must be handled. The following flags are defined:

V (Vendor-Specific)

When set, this flag indicates that the optional Vendor-ID field is present and the AVP Code belongs to the vendor. The Vendor-ID field ensures that there are no collisions between vendor-specified AVP Codes.

M (Mandatory)

If set, the receiver must be able to understand the AVP's semantics and content. If the receiver cannot, it sends an appropriate error message, unless the mandatory

Table 2.4 Base AVPs.

Attribute Name	AVP Code	Data Type	AVP Flag
Acct-Interim-Interval	85	Unsigned32	M
Accounting-Realtime-Required	483	Enumerated	M
Acct-Multi-Session-Id	50	UTF8String	M
Accounting-Record-Number	485	Unsigned32	M
Accounting-Record-Type	480	Enumerated	M
Acct-Session-Id	44	OctetString	M
Accounting-Sub-Session-Id	287	Unsigned64	M
Acct-Application-Id	259	Unsigned32	M
Auth-Application-Id	258	Unsigned32	M
Auth-Request-Type	274	Enumerated	M
Authorization-Lifetime	291	Unsigned32	M
Auth-Grace-Period	276	Unsigned32	M
Auth-Session-State	277	Enumerated	M
Re-Auth-Request-Type	285	Enumerated	M
Class	25	OctetString	M
Destination-Host	293	DiameterIdentity	M
Destination-Realm	283	DiameterIdentity	M
Disconnect-Cause	273	Enumerated	M
Error-Message	281	UTF8String	
Error-Reporting-Host	294	DiameterIdentity	
Event-Timestamp	55	Time	M
Experimental-Result	297	Grouped	M
Experimental-Result-Code	298	Unsigned32	M
Failed-AVP	279	Grouped	M

AVP is embedded within an optional AVP of type Grouped (Table 2.6), or if the receiver is a Diameter relay or redirect agent.

AVPs with the "M" bit cleared are informational, and a receiver that doesn't support the AVP or its value may ignore it. If an AVP of data type Grouped does not have the "M" bit set, a receiver may ignore the AVP, even if one or more AVPs within the Grouped AVP does have the "M" bit set.

P (End-to-end security)

In RFC 3588, the "P" bit was used to indicate that the AVP needed to be encrypted for end-to-end security, but since such a security mechanism has not yet been defined, the "P" bit should be set to 0, and be considered reserved for future use. RFC 6733 deprecated the use of "P" bit.

r (Reserved)

Reserved for future use.

Table 2.5 Base AVPs, continued.

Attribute Name	AVP Code	Data Type	AVP Flag
Firmware-Revision	267	Unsigned32	
Host-IP-Address	257	Address	M
Inband-Security-Id	299	Unsigned32	M
Multi-Round-Time-Out	272	Unsigned32	M
Origin-Host	264	DiameterIdentity	M
Origin-Realm	296	DiameterIdentity	M
Origin-State-Id	278	Unsigned32	M
Product-Name	269	UTF8String	
Proxy-Host	280	DiameterIdentity	M
Proxy-Info	284	Grouped	M
Proxy-State	33	OctetString	M
Redirect-Host	292	DiameterURI	M
Redirect-Host-Usage	261	Enumerated	M
Redirect-Max-Cache-Time	262	Unsigned32	M
Result-Code	268	Unsigned32	M
Route-Record	282	DiameterIdentity	M
Session-Id	263	UTF8String	M
Session-Timeout	27	Unsigned32	M
Session-Binding	270	Unsigned32	M
Session-Server-Failover	271	Enumerated	M
Supported-Vendor-Id	265	Unsigned32	M
Termination-Cause	295	Enumerated	M
User-Name	1	UTF8String	M
Vendor-Id	266	Unsigned32	M
Vendor-Specific-Application-Id	260	Grouped	M

Figure 2.6 AVP header format.

Table 2.6 Basic AVP data formats.

Data Format	Data Field Contains	AVP Length
OctetString	Arbitrary data	Variable length, padded at the end with zero-valued bytes to align on a 32-bit boundary
Integer32	32-bit signed value	12 (16)
Integer64	64-bit signed value	16 (20)
Unsigned32	32-bit unsigned value	12 (16)
Unsigned64	64-bit unsigned value	16 (20)
Float32	This 32-bit value represents floating point values of single precision	12 (16)
Float64	This 64-bit value represents floating point values of double precision	16 (20)
Grouped	Sequence of AVPs, including their headers and padding, that are concatenated in the order in which they are specified. A Grouped AVP may contain AVPs that are also of type Grouped.	8 (12) plus the total length of included AVPs

Parentheses indicate length if the V bit is enabled.

AVP Length

The AVP Length field indicates the number of octets in the AVP, including the AVP Code field, AVP Length field, AVP Flags field, Vendor-ID field (if present), and AVP Data field. If a message is received with an invalid attribute length, the receiver rejects the message.

Vendor-ID (optional)

The AVP header can contain the optional Vendor-ID field, which identifies a specific vendor. The Vendor-ID field is present if the "V" bit is set in the AVP Flags field. This optional field contains a value assigned to the vendor by IANA. Any vendors or standardization organizations implementing their own proprietary Diameter AVPs provide their Vendor-ID along with their privately managed AVP Code, so that they do not conflict with any other vendor's proprietary AVP(s) or with future IETF AVPs. A Vendor-ID value of zero (0) indicates that the AVP values have been specified by the IETF.

Data

The Data field contains the information for the attribute and can be zero or more octets in length. The format and length of the Data field is specified by its AVP Code and AVP Length fields. The Data field's format can be one of the base data types found in Table 2.6 or it can be a data type derived from the base data types.

Each AVP of type OctetString, which holds arbitrary data, is padded to align on a 32-bit boundary, while other AVP types align naturally. A number of zero-valued bytes are added to the end of the AVP data field until a word boundary is reached. The length of the padding is not reflected in the AVP Length field.

2.6.4 Derived AVP Data Formats

The following are common derived AVP data formats. If a derived data format is based on the OctetString data format, then the AVPs of that type are padded to align on a 32-bit boundary. Diameter applications may define their own derived formats.

Address

The Address data format is derived from the OctetString data format. It identifies the type of address it contains, for example a 32-bit (IPv4) or 128-bit (IPv6) address, and then provides the content.

Time

The Time data format is derived from the OctetString data format and follows the format of the Network Time Protocol (NTP) timestamp data type, which represents the number of seconds since 0h on 1 January 1900 UTC [9].

UTF8String

The UTF8String is derived from the OctetString data format. This is a human-readable string.

DiameterIdentity

The DiameterIdentity format is derived from the OctetString data format. The contents are either a realm, which helps determine whether a message can be handled locally or needs to be routed, or a fully qualified domain name, which identifies a Diameter node.

DiameterURI

The DiameterURI format follows the Uniform Resource Identifiers syntax rules specified in RFC 3986 [10]. A URI specifies at a minimum the fully qualified domain name, and may contain the port, transport or protocol.

Enumerated

The Enumerated data format is derived from the Integer32 data format. An AVP with this data type contains a list of valid values and their interpretation.

IpFilterRule

The IpFilterRule format is derived from the OctetString data format, and its syntax is a subset of the syntax for the FreeBSD firewall, ipfw. AVPs using the IpFilterRule format can specify how to categorize traffic and filter a user's packets. Although defined in the base specification, this format is not used by any base AVPs. The IpFilterRule format is used within the Diameter Network Access Application defined in RFC 7155 [11].

2.7 Diameter Sessions

A session, which is related to the service provided to an end user, is a sequence of related Diameter messages and is a logical concept at the application layer. Sessions are identified with the same Session-Id AVP in each message. Each message also carries the Application-Id of the application. A session is routed only through authorized nodes that have advertised support for the application. A sub-session is a distinct service (e.g., Quality of Service, data characteristics) provided to a session, and is tracked with the Accounting-Sub-Session-Id AVP. Sub-sessions may occur concurrently with each other or they may happen serially.

Note that there is no relationship between transport connections and sessions – multiple sessions can be carried on a single transport connection.

2.8 Transaction Results

The `Result-Code` AVP in the Diameter answer message indicates success or failure of the request, or that more information is forthcoming. The `Result-Code` AVP indicates one of five result categories based on the thousands digit in decimal notation:

- 1xxx – Informational
- 2xxx – Success
- 3xxx – Protocol errors
- 4xxx – Transient errors, a temporary kind of application error
- 5xxx – Permanent failure, a non-recoverable kind of application error

If there are multiple errors, the Diameter node reports only the first error it encountered. Applications may define application-specific errors in addition to using the base protocol errors.

2.8.1 Successful Transactions

The base specification defines two success codes: `DIAMETER_SUCCESS` (2001), which indicates the request was successfully completed, and `DIAMETER_LIMITED_SUCCESS` (2002), which also indicates the successful completion of a request, but that further application processing is required in order to provide service to the user.

The only informational Result-Code specified by the IETF at time of writing is `DIAMETER_MULTI_ROUND_AUTH` (1001), which informs the client that the authentication mechanism being used requires multiple round trips, and that it should send another request within the time specified in the `Multi-Round-Time-Out` AVP before access is granted.

2.8.2 Protocol Errors

Protocol errors occur at the base protocol level. An example is a message routing error. A Diameter node indicates a protocol error by setting the "E" bit in its answer message. The answer message also has a different CCF than a regular answer (Figure 2.7).

A proxy may attempt to fix the protocol error when it receives the answer message. The following `Result-Code` values indicate protocol errors:

- `3001 DIAMETER_COMMAND_UNSUPPORTED`: The node does not support the request's Command Code.
- `3002 DIAMETER_UNABLE_TO_DELIVER`: The agent or proxy could not find a server to support the required application or the request had a `Destination-Host` AVP but not a `Destination-Realm` AVP.
- `3003 DIAMETER_REALM_NOT_SERVED`: The node does not recognize the destination realm.
- `3004 DIAMETER_TOO_BUSY`: The requested server cannot provide service. The client should send the message to another peer.

```
<answer-message> ::= < Diameter Header: code, ERR [, PXY] >
                 0*1< Session-Id >
                    { Origin-Host }
                    { Origin-Realm }
                    { Result-Code }
                    [ Origin-State-Id ]
                    [ Error-Message ]
                    [ Error-Reporting-Host ]
                    [ Failed-AVP ]
                    [ Experimental-Result ]
                  * [ Proxy-Info ]
                  * [ AVP ]
```

Figure 2.7 An example of the Command Code Format for a Diameter answer command that reports an error.

- 3005 DIAMETER_LOOP_DETECTED: The node found its own identity in a Route-Record AVP while attempting to send the message.
- 3006 DIAMETER_REDIRECT_INDICATION: The Diameter agent requests that the client contact another agent or the server directly using information in the included Redirect-Host AVP.
- 3007 DIAMETER_APPLICATION_UNSUPPORTED: The node does not support the given application.
- 3008 DIAMETER_INVALID_HDR_BITS: The bits in the request's Diameter header were set either invalidly or inconsistently.
- 3009 DIAMETER_INVALID_AVP_BITS: The AVP included flag bits set either invalidly or inconsistently.
- 3010 DIAMETER_UNKNOWN_PEER: The node received a CER from an unknown peer and is not configured to interact with them. When a peer is unknown, the node can also just silently discard the CER without answering.

Some of these errors are covered in greater detail in Section 4.4.2.

Figure 2.8 shows an example of a message forwarded by a Diameter relay. Relay 2 determines that it cannot forward the request to the server, so it returns an answer message with the "E" bit set and the Result-Code AVP set to DIAMETER_UNABLE_TO_DELIVER. Since this a protocol error, Relay 1 takes special action, and attempts to route the message through its alternate Relay 3.

Note: The term "home" or "local" refers to the administrative domain with which the user maintains an account relationship. A home server can process a Diameter request directly and does not need to route the request elsewhere.

2.8.3 Transient Failures

Transient failures occur when the server could not satisfy the request at the time it had received it, but it could possibly handle it successfully in the future. When the server responds with a transient failure it does not set the "E" bit.

- 4001 DIAMETER_AUTHENTICATION_REJECTED: The server returns this if the authentication process for the user failed, probably due to incorrect credentials provided by the user. The client should request new credentials from the user.

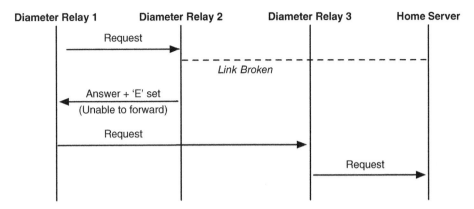

Figure 2.8 Protocol error call flow.

- `4002 DIAMETER_OUT_OF_SPACE`: The server was unable to store an accounting request's information due to a lack of space.
- `4003 ELECTION_LOST`: In the scenario where two nodes initiate a connection to each other at the same time, a process known as an *election* is invoked to determine which node keeps the connection up. With this error message, the peer indicates that it lost the election. For more information, see Section 3.5.1.

2.8.4 Permanent Failures

Permanent failures occur when the request failed and the peer should not attempt the request again. The "E" bit should not be set in the answer, however some permanent failure errors defined by a Diameter application may set the "E" bit.

- `5001 DIAMETER_AVP_UNSUPPORTED`: The peer does not recognize or support an AVP that was marked with the "M" bit, and indicates which AVP(s) are unsupported by including them in the `Failed-AVP` AVP in the answer.
- `5002 DIAMETER_UNKNOWN_SESSION_ID`: The request contained an unknown or inactive `Session-Id`.
- `5003 DIAMETER_AUTHORIZATION_REJECTED`: The server could not authorize the user or the user was not allowed to use the service requested. The server also uses this result code if the server found an untrusted realm in the `Route-Record` AVP.
- `5004 DIAMETER_INVALID_AVP_VALUE`: The request contained an AVP with an invalid value. The node indicates which AVP(s) had invalid values by including them in the `Failed-AVP` AVP(s) in the answer.
- `5005 DIAMETER_MISSING_AVP`: The command is missing a mandatory AVP or the `Vendor-Specific-Application-Id` AVP did not contain either an `Auth-Application-Id` AVP or an `Acct-Application-Id` AVP. The peer includes an example of the missing AVP along with the `Vendor-Id` if applicable in the `Failed-AVP` AVP. If a `Vendor-Specific-Application-Id` was incomplete, then the peer includes a `Failed-AVP` AVP with an example of an `Auth-Application-Id` AVP and an `Acct-Application-Id` AVP.
- `5006 DIAMETER_RESOURCES_EXCEEDED`: The server could not authorize the user because the user had already expended its allotted resources.

- 5007 DIAMETER_CONTRADICTING_AVPS: The request contained AVPs that contradicted each other. The server includes these contradicting AVPs in the Failed-AVP AVP.
- 5008 DIAMETER_AVP_NOT_ALLOWED: The request contained an AVP, identified by the Failed-AVP AVP, that is not allowed by the command.
- 5009 DIAMETER_AVP_OCCURS_TOO_MANY_TIMES: The request contained too many instances of an AVP. The peer includes a copy of the first instance of the AVP that exceeded the maximum number of occurrences in the Failed-AVP AVP.
- 5010 DIAMETER_NO_COMMON_APPLICATION: The peer has no Diameter applications in common with those specified in the CER.
- 5011 DIAMETER_UNSUPPORTED_VERSION: The peer does not support the version number given in the request.
- 5012 DIAMETER_UNABLE_TO_COMPLY: The peer has rejected the request for unspecified reasons.
- 5013 DIAMETER_INVALID_BIT_IN_HEADER: The request had the Reserved bit set to one (1) or Diameter header bits are set incorrectly.
- 5014 DIAMETER_INVALID_AVP_LENGTH: The request contained an AVP with an invalid length. The node includes the AVP and a zero-filled payload of the minimum required length for the data type in the Failed-AVP AVP. If the node could not decode a truncated AVP, the node includes the AVP header and pads it with zeros up to the minimum AVP header length.
- 5015 DIAMETER_INVALID_MESSAGE_LENGTH: The request's length was invalid.
- 5016 DIAMETER_INVALID_AVP_BIT_COMBO: The request contained an AVP with an invalid value in the AVP Flags field. The node includes the AVP in the Failed-AVP AVP.
- 5017 DIAMETER_NO_COMMON_SECURITY: The peer has no security mechanisms in common with those specified in the CER. This may occur when interacting with 3588 implementations that negotiate security with the Inband-Security-Id AVP.

Figure 2.9 shows a Diameter message causing an application error. An application error occurs when there is a problem with an application-specified function, like a user authentication error, and can be either a transient or a permanent failure. When an application error occurs, the Diameter node reporting the error sends an answer

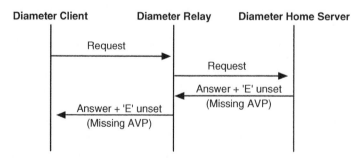

Figure 2.9 Application error call flow.

with the appropriate `Result-Code` AVP. Since application errors do not require any proxy or relay agent involvement, the message is simply forwarded back to the request originator.

If the answer itself contains errors, the recipient can terminate the related session by sending a Session-Termination-Request or Abort-Session-Request message and indicating the cause of the error in the `Termination-Cause` AVP. An application could also send an application-specific request to signal the error if no state is maintained.

2.9 Diameter Agents

2.9.1 Saving State

In order to provide failover support, all agents maintain a transaction state by saving the request's unique identifier, known as a Hop-by-Hop Identifier, which is designed to match answers with requests. If an agent forwards a request, it replaces the request's Hop-by-Hop Identifier with its own locally unique identifier. When the agent receives an answer to the forwarded request, it restores the original value and sends the answer toward the requester. It then releases the transaction state that it was maintaining.

However, maintaining the transaction state doesn't mean that an agent is necessarily stateful. Only agents that maintain a session state are considered stateful. A session state is the tracking of all authorized, active sessions. The agent considers a session active until it is notified otherwise or the session expires. This expiration information is provided in the Diameter message via the `Session-Timeout` AVP.

2.9.2 Redirect Agents

A redirect agent informs a client how to communicate directly with a server or other agents, and then it removes itself from the communication path. Figure 2.10 shows a client contacting a redirect agent (step 1). The redirect agent sends an answer message (step 2) with the "E" bit set, while maintaining the Hop-by-Hop Identifier in the header. It includes `DIAMETER_REDIRECT_INDICATION` in the `Result-Code` AVP, and how to contact the server by including the server's DiameterURI in one or more `Redirect-Host` AVPs. The client stores this information and contacts the server directly (step 3). Redirect agents are useful in networks where the routing configuration needs to be centralized. Details are covered in Section 3.4.2.

Redirect agents, since they simply return information to clients and do not modify messages, are capable of handling any Diameter application or message type, although they may be configured to redirect messages of certain types only. Since they do not receive answers to their messages, they do not maintain a session state. Redirect agents advertise their service with the Relay Application-Id, 4294967295.

2.9.3 Relay Agents

Relay agents are responsible for finding a server that supports the application of a particular message and then forwarding that message onward. Relay agents do not originate

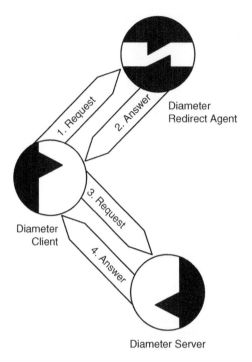

Figure 2.10 Diameter redirection.

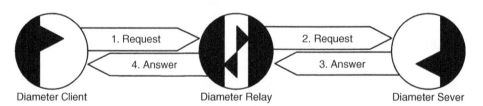

Figure 2.11 Diameter relay.

messages, nor understand message semantics, nor make policy decisions; however, they can handle any Diameter application or message type. A relay agent only modifies messages by inserting routing information, and does not keep the session state. Relay agents advertise their service with the Relay Application-Id, 4294967295.

Figure 2.11 shows a client sending a message to a relay agent (step 1). The relay agent forwards the message (steps 2 and 4) after updating the Hop-by-Hop Identifier and appending a `Route-Record` AVP, which contains the identity of the peer from which the agent received the request.

A relay agent can act as request aggregator for multiple clients and send those requests to other realms. This reduces the configuration burdens on both clients and servers. Diameter clients don't need to be configured with the security information required to communicate with Diameter servers in other realms, and Diameter servers don't need to be reconfigured when clients are added, removed, or changed.

2.9.4 Proxy Agents

Like relay agents, proxy agents forward requests and responses; however, proxy agents provide more services since they can enforce resource usage policies and can provide access control and provisioning. For example, a proxy may inform a server that a session is terminated with a Session-Termination-Request message. Because proxy agents make policy decisions, they must understand message semantics and only support those applications for which they have been designed. Thus, a Diameter proxy is referred to as a Diameter *Application* proxy, where *Application* is the application that it supports.

2.9.5 Translation Agents

A Diameter translation agent translates between another AAA protocol, such as RADIUS, and Diameter. Translation agents can only translate requests and answers of applications that they recognize. Since Diameter supports long-lived sessions, translation agents maintain the session state. A translation agent can provide a gateway to Diameter infrastructure, which allows legacy systems to migrate at a slower pace.

References

1 IANA. Radius Types: Authentication, Authorization, and Accounting (AAA) Parameters, Jan. 2016. http://www.iana.org/assignments/aaa-parameters/aaa-parameters.xhtml.
2 D. Sun, P. McCann, H. Tschofenig, T. Tsou, A. Doria, and G. Zorn. Diameter Quality-of-Service Application. RFC 5866, Internet Engineering Task Force, May 2010.
3 J. Postel. Transmission Control Protocol. RFC 0793, Internet Engineering Task Force, Sept. 1981.
4 R. Stewart. Stream Control Transmission Protocol. RFC 4960, Internet Engineering Task Force, Sept. 2007.
5 T. Dierks and E. Rescorla. The Transport Layer Security (TLS) Protocol Version 1.2. RFC 5246, Internet Engineering Task Force, Aug. 2008.
6 M. Tuexen, R. Seggelmann, and E. Rescorla. Datagram Transport Layer Security (DTLS) for Stream Control Transmission Protocol (SCTP). RFC 6083, Internet Engineering Task Force, Jan. 2011.
7 S. Kent and K. Seo. Security Architecture for the Internet Protocol. RFC 4301, Internet Engineering Task Force, Dec. 2005.
8 D. Crocker and P. Overell. Augmented BNF for Syntax Specifications: ABNF. RFC 5234, Internet Engineering Task Force, Jan. 2008.
9 D. Mills, J. Martin, J. Burbank, and W. Kasch. Network Time Protocol Version 4: Protocol and Algorithms Specification. RFC 5905, Internet Engineering Task Force, June 2010.
10 T. Berners-Lee, R. Fielding, and L. Masinter. Uniform Resource Identifier (URI): Generic Syntax. RFC 3986, Internet Engineering Task Force, Jan. 2005.
11 G. Zorn. Diameter Network Access Server Application. RFC 7155, Internet Engineering Task Force, Apr. 2014.

3

Communication between Neighboring Peers

3.1 Introduction

This chapter describes in detail the principles of both transport and Diameter-level connectivity between two adjacent Diameter nodes, known as peers. This chapter covers how the peers discover each other, how the peers connect and maintain connections, and how transport failures are handled. The chapter also covers advanced transport topics such as multi-homed connections, head-of-line blocking, and multiple connection instances.

3.2 Peer Connections and Diameter Sessions

Two Diameter nodes that have a direct transport protocol connection between each other, known as a *peer connection*, are considered Diameter peers. The peer connection does not need to be directly connected from the IP point of view; there may be zero or more IP routers between the Diameter peers. Figure 3.1 shows a simple network topology where two peers are connected over an arbitrary IP network.

The two peers may also establish a *Diameter user session* between each other. However, it is not required that a peer connection map directly to a Diameter session. Figure 3.2 shows a simplified network topology where Diameter nodes A and B have a Diameter user session that consists of multiple peer connections through an intermediating Diameter agent C.

It should be understood that Diameter decouples the peer connection and the Diameter user session. The lifetime and presence of Diameter user sessions are independent of the specific peer connections. On the other hand, a peer connection is directly dependent on the underlying transport connection.

3.3 The DiameterIdentity

Each Diameter node has a DiameterIdentity, which is a fully qualified domain name (FQDN). A node's DiameterIdentity must be unique within the host where the Diameter

Diameter: New Generation AAA Protocol – Design, Practice, and Applications, First Edition.
Hannes Tschofenig, Sébastien Decugis, Jean Mahoney and Jouni Korhonen.
© 2019 John Wiley & Sons Ltd. Published 2019 by John Wiley & Sons Ltd.

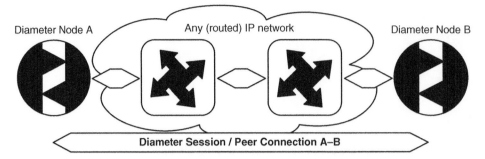

Figure 3.1 Diameter and transport connection between two peers.

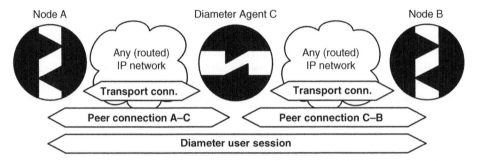

Figure 3.2 Diameter user sessions and peer connections.

node is located. If the node is reachable via the Internet, its DiameterIdentity should be globally unique as well. It is important that the DiameterIdentity is unique because of its role in Diameter connection management. A node uses a peer's DiameterIdentity as the peer table lookup key (see Section 3.7) and tracks the peer state machine (see Section 3.8.2). A node may also use a DiameterIdentity to discover peers dynamically (see Section 3.4.2).

A DiameterIdentity maps to one or more IP addresses of a Diameter node. A Diameter node may be:

- single-homed and single addressed,
- single-homed and multiple addressed, or
- multi-homed and multiple addressed.

An IP address belongs to one and only one DiameterIdentity, that is, to a single Diameter node. A networking node may host multiple Diameter nodes and thus multiple DiameterIdentities, for example when every Diameter node is a separate virtual machine or a process. Regardless, the requirement holds that an IP address always maps to a single DiameterIdentity and each Diameter node has a single unique DiameterIdentity. We will discuss the role of multiple IP addresses later in the context of *multiple connection instances* in Section 3.9.4.

3.4   Peer Discovery

Peer discovery refers to the mechanism that a Diameter node uses to discover other peers to which it should attempt to establish connections with. There are two types of peer discovery: static discovery, in which connection information for adjacent peers is configured into the Diameter node, and dynamic discovery, in which a Diameter node uses Domain Name System (DNS) to find its adjacent peers.

3.4.1   Static Discovery

Static peer discovery is the traditional way for AAA systems to find their adjacent peers. Diameter is no exception. A typical way for a Diameter node to learn its peers, their IP addresses or FQDNs, and the rest of configuration information (such as the security-related parameters and associated realms) is to read a configuration file or a database. As a result, the peer configuration and especially the discovery are static in nature.

Once the peer configuration is known, Diameter node's peer and routing tables are configured appropriately. Apart from the application identifier information, the peer discovery is now completed and remains static until new peer configuration information is added, changed, or removed through administrative actions.

3.4.1.1   Static Discovery in `freeDiameter`

`freeDiameter` statically discovers its peers by reading its configuration file. By default, `freeDiameter` will reject connections from unknown peers. An example of peer configuration can be found in the *freeDiameter-1.conf* file found in the `/src/doc/single_host/` directory. At the end of the file is the following line:

```
ConnectPeer = "peer2.localdomain" { ConnectTo = "127.0.0.1";
                                    No_TLS; port = 30868; };
```

Where `peer2.localdomain` is the DiameterIdentity of the peer. Note that the configured DiameterIdentity must match the information received inside the CEA or the connection will be aborted. The `No_TLS` parameter tells `freeDiameter` to assume transparent security instead of TLS.

The `ConnectPeer` configuration has the following general format:

```
ConnectPeer = "<diameterid>" [ { <parameter>; <parameter>; ...} ] ;
```

The following parameters can be specified in the peer's parameter list, separated by semicolons:

- IPv4 or IPv6 address of the peer: `ConnectTo = "<IP_address>"`
- Transport protocol information: `No_TCP`, `No_SCTP`, `No_IP` (use only IPv6 addresses), `No_IPv6` (use only IPv4 addresses), `Prefer_TCP`, `TLS_old_method` (use the RFC 3588 method for TLS protection, where TLS is negotiated after CER/CEA exchange is completed), `No_TLS`. Note that DTLS is not yet supported.

- Port information: `Port = <port_number>`
- Tc Timer setting, which specifies how often to attempt to make a transport connection when one doesn't exist: `TcTimer = <seconds>`. The default value is 30 seconds. For more information about the Tc Timer, see Section 3.8.1.
- Tw Timer setting, which is used to trigger timeouts at the Diameter application level: `TwTimer = <seconds>`. The default value is 30 seconds. For more information about the Tw Timer, see Section 3.8.1.
- GNU TLS priority string, which allows you to configure the behavior of GNUTLS key exchanges: `TLS_Prio = "<priority>"`. The default value is `NORMAL`. See the *gnutls_priority_init* function documentation for more information.
- Realm, if the peer does not advertise the given realm, reject the connection: `Realm = "<realm>"`

If a parameter is not specified, its default is used. Here are some more examples of peer configuration:

```
ConnectPeer = "aaa.wide.ad.jp";
ConnectPeer = "old.diameter.serv" { TcTimer = 60; TLS_old_method;
                                     No_SCTP; Port=3868; } ;
```

3.4.2 Dynamic Discovery

When no matching statically configured peers exist or if none of its existing peers responds, a Diameter node can discover adjacent peers dynamically by consulting the DNS. The DNS is a system that primarily translates (fully qualified) domain names into IP addresses.[1]

The benefits of the dynamic discovery are obvious. Dynamic discovery eases the management of the Diameter nodes since the peer and routing information does not need to be known before initial deployment. Adding new peers or destination realms ideally requires no additional configuration. The administrator of a realm populates the realm's master DNS servers with the information of the Diameter nodes that he or she wants to be discoverable. There may be multiple provisioned nodes per realm with the same or different set of supported Diameter applications, transport protocols, and security profiles. If security is used, then appropriate certificates for all peers with which the Diameter node may attempt to establish a connection must be installed and available.

It is important to understand that DNS-based dynamic discovery is different from resolving a Diameter peer's DiameterIdentity to an IP address for transport connection establishment, although at the end of the discovery process DiameterIdentity to IP address resolving also takes place. When looking up peers in DNS, the Diameter node uses the destination realm, the preferred/supported transport, and security mechanisms, rather than using the peer's DiameterIdentity.

Although it may appear similar, dynamic discovery is not the same as dynamic AAA routing. The discovery helps in finding peers, but there may be a number of intermediate Diameter agents between the discovery initiator and the discovered peer that must be traversed. The details of different deployments are discussed in more detail in Chapter 4.

---

1 RFC 3588 had also included a dynamic discovery mechanism based on Service Location Protocol version 2 (SLPv2) [1], but SLPv2 was rarely used in Diameter deployments and was deprecated by RFC 6733.

```
;;          order pref flags service        regexp  replacement
IN NAPTR 25    50    "s"   "AAA+D2S"      ""      _diameter._sctp
                                                   .example.com
IN NAPTR 100   50    "s"   "AAA+D2T"      ""      _aaa._tcp.example.com
```

Figure 3.3 An example of legacy NAPTR records.

The concepts of DNS span many IETF specifications and have evolved over the years. The following paragraphs provide an overview of how Diameter uses DNS.

RFC 3588 specified the use of an older form of Naming Authority Pointer (NAPTR) [2] resource records (RRs), which were then used to retrieve Service Location (SRV) RRs [3], which provide locations of specific nodes based on the transports the nodes support. An example of legacy NAPTR records for Diameter can be found in Figure 3.3.

These records tell the querying node that SCTP and TCP are supported. The lower order number given for the SCTP entry indicates that SCTP is preferred. The flags column tells the querying node what the replacement column contains. If "A" is present in the flags field, then the replacement value contains a domain name to be used to look up an address record (known as an A or AAAA record [4, 5]). If the flags field is empty, then the querying node should use the replacement value in a new NAPTR query. Note that this recursion is error prone and has since been deprecated in newer DNS specifications.

The "s" flag tells the querying node to use the replacement value, in this case _diameter._sctp.example.com, in an SRV query to determine the location of the service. Figure 3.4 shows the results of the SRV query.

The querying node then uses the domain name found in the SRV RR target field to do an address record lookup. Once an IPv4 or IPv6 address is retrieved, the node has the information necessary to attempt a transport connection with the discovered peer. However, note that these DNS records do not provide information on supported Diameter applications. If DNS returned multiple Diameter nodes, the querying node has to iterate through these Diameter nodes to find a node that supports the same applications. This means establishing a peer connection, running the CER/CEA exchange, and then possibly discarding the connection if no desired applications are found. In a large Diameter network, this could add a considerable amount of unnecessary connection attempt overhead and delay during the discovery phase.

These concerns resulted in the development of RFC 6408 [6], which introduced a dynamic discovery procedure based on the Dynamic Delegation Discovery Service (DDDS) [7] and Straightforward-NAPTR (S-NAPTR) [8], which added Diameter Application-Id information to the S-NAPTR records.

Figure 3.5 shows an example of S-NAPTR records that show support for particular Diameter applications. In these S-NAPTR records, Diameter applications are indicated in the service field by using a naming convention aaa+ap<Application-Id>. In the example, the server supports both the NASREQ application (Application-Id 1) and the Credit Control application (Application-Id 4).

```
;;          Priority Weight Port   Target
IN SRV  0          1      5060   server1.example.com
IN SRV  0          2      5060   server2.example.com
```

Figure 3.4 Results of an SRV query.

```
;;             order pref flags service                      regexp replacement
IN NAPTR  50    50    "s"   "aaa:diameter.sctp"       ""   _diameter._sctp
                                                           .ex1.example.com
IN NAPTR  50    50    "s"   "aaa+ap1:diameter.sctp"  ""   _diameter._sctp
                                                           .ex1.example.com
IN NAPTR  50    50    "s"   "aaa+ap4:diameter.sctp"  ""   _diameter._sctp
                                                           .ex1.example.com
```

Figure 3.5 S-NAPTR records showing support for Diameter applications.

```
;;          order pref flags service                    regexp replacement
IN NAPTR 50    50    "A"   "aaa:diameter.dtls.sctp" ""       "node.example.com"
```

Figure 3.6 An RFC 6733 style S-NAPTR record.

```
;; provisioned under the "$ORIGIN example.com."
;;
;;          order pref flags service                    regexp replacement
IN NAPTR 50    50    "S"   "aaa+ap8:diameter.tls.tcp" ""    "_diameters._tcp.mip
                                                            .example.com"
IN NAPTR 50    50    "S"   "aaa+ap8:diameter.tcp"       ""  "_diameter._tcp.mip
                                                            .example.com"

;;                            Priority  Weight  Port    Target
_diameters._tcp.mip IN SRV    0         1       5868    ha.mip.example.com
_diameter._tcp.mip IN SRV     0         1       3868    ha.mip.example.com

;;
ha.mip             IN  A      192.0.2.1
ha.mip             IN  AAAA   2001:bd8:dead:beef::c000::201
```

Figure 3.7 An example of DNS RRs for realm "example.com".

RFC 6733 aligned with the NAPTR DDDS application-based, dynamic-discovery mechanism defined by RFC 6408. Also, although RFC 6733 states that using DNS for dynamic discovery is mandatory to implement, it is optional to deploy. RFC 6733 also does not mandate the use of the application identifier part of the RFC 6408 algorithm, so Figure 3.6 is also a valid example of an RFC 6733 style S-NAPTR record.

Figure 3.7 shows an example of DNS RRs for the realm "example.com". The DNS configuration shows support for TLS (or no TLS) transport over TCP, the use of SRV indirection, and the Mobile IPv6 Auth Diameter application (Application-Id 8 [9]). Note that _diameters in the replacement field indicates a secure version of the transport protocol.

Legacy applications still use RFC 3588 style NAPTRs and therefore the DNS administrator for a Diameter deployment should also provision NAPTR RRs in RFC 3588 style for backward compatibility.

The following algorithm example for dynamic discovery follows the original description in RFC 6733 and the modifications introduced by RFC 6408:

1. Before starting the procedure, the querying node must know the realm to contact. The querying node may extract the realm information, for example from the realm part of the User-Name AVP value found in the request message. Details of using information from the User-Name AVP can be found in Section 4.3.1.

2. The Diameter node performs a NAPTR query for a service in the particular realm (e.g., "example.com").
3. If the returned NAPTR RRs contain service fields formatted as `aaa+ap<X>:<Y>`, where `<X>` is a Diameter Application-Id and `<Y>` is one of the following, indicating the supported transport protocol:
   - `diameter.tcp`
   - `diameter.sctp`
   - `diameter.dtls`
   - `diameter.tls.tcp`

   then the queried realm supports the DDDS NAPTR application as defined in RFC 6408. An example service field is the following: `aaa+ap1:diameter.tcp`.

   (a) If the `<X>` contains the desired Application-Id and the `<Y>` contains a desired transport protocol and security combination, the resolver resolves the replacement field to a target host using the lookup mechanism defined by the flags field.
   (b) If neither `<X>` nor `<Y>` contain desired values, then the querying node discards the current entry and examines the next returned NAPTR RR, if such exists.
4. If the returned NAPTR RRs contain service fields formatted as `aaa+ap<X>`, the querying node resolves the replacement field to a target host using the lookup mechanism defined by the flags field [8]. All possible transport protocols are tried in the following order: TLS, followed by DTLS, then by TCP, and finally by SCTP.
5. If the returned NAPTR RRs contain service fields formatted as `aaa:<Y>`, where `<Y>` indicates the supported transport protocol, then the queried realm supports the simplified DDDS NAPTR application given in RFC 6733.
   (a) If the `<Y>` contains a supported and preferred transport protocol and security combination, the querying node resolves the replacement field to a target host using the lookup mechanism defined by the flags field.
6. If the returned NAPTR RRs contain service fields formatted as either `aaa+D2T` for TCP or `aaa+D2S` for SCTP,[2] the realm supports dynamic discovery defined in RFC 3588, and the querying node should fall back to using legacy NAPTR [2].
7. If none of the above returned a positive result, the queried realm does not support NAPTR-based Diameter peer discovery. The querying node requests SRV RRs directly from the realm using the same transport protocol prioritization as described above.
8. If the SRV query fails, then the querying node may try to resolve address records (A/AAAA RRs) for the target hostname specified in the SRV RRs and following the rules laid out in [3, 6, 10].
9. If all of the above fails, the querying node returns an error to the Diameter application implementation.

3.4.2.1 Dynamic Discovery and DiameterURI

Diameter nodes can also be discovered by inspecting AVPs, for instance the `Redirect-Host` AVP, that use the derived datatype *DiameterURI*. A node may receive one or more `Redirect-Host` AVPs in an answer message from a redirect agent to inform the requesting node to send the request to a different Diameter node.

---

2 Although RFC 3588 specified the creation of an IANA registry for these service fields, a registry was never created.

The DiameterURI datatype follows the Uniform Resource Identifier (URI) syntax [11] and can capture peer node's port number, FQDN, transport protocol, and security. Below is an example syntax of the DiameterURI with and without secure transport. Note that we are showing just the high-level URI syntax and not the actual Augmented Backus–Naur Form (ABNF) [12] details for the URI. For the ABNF details, see RFC 6733.

A Diameter peer that does not support secure transport is represented by a DiameterURI starting with `aaa`:

```
"aaa://" FQDN [port] [transport] [protocol]
```

A Diameter peer that supports secure transport is represented by a DiameterURI starting with `aaas`:

```
"aaas://" FQDN [port] [transport] [protocol]
```

The example DiameterURI below is one of the DNS examples with TCP transport and TLS security from Section 3.4.2:

```
aaas://ha.mip.example.com:5868;transport=tcp;protocol=diameter
```

There are a few notes to make about the DiameterURI. First, the `port` parameter is used to specify non-standard port numbers for Diameter. If the parameter is absent, default ports apply (TCP and SCTP 3868, TLS/TCP and DTLS/SCTP 5868).

Second, the `transport` parameter allows UDP to be specified as the transport protocol, even though it is not possible to use UDP with Diameter. However, closer inspection of the DiameterURI reveals that it also allows AAA protocols other than Diameter to be specified in the `protocol` parameter, namely both RADIUS [13] and TACACS+ [14]. Obviously, in Diameter deployments only those transports and protocols supported by Diameter can be used.

Finally, coming back to the dynamic discovery, the FQDN given in the URI needs to be resolved using DNS. In this case, a normal A/AAAA query is used [4, 5]. Furthermore, differing from the general URI specification, the DiameterURI always contains an FQDN, and never an IP literal.

3.4.2.2 DNS Further Reading

As mentioned earlier, DNS is a broad topic covered in many IETF RFCs. For more information, consult the RFCs mentioned in this section: RFC 6408 [6], which covers S-NAPTR usage for Diameter, RFC 3958 [8], which covers S-NAPTRs and Dynamic Delegation Discovery Service, and RFC 2783 [3], which covers SRV records. RFC 1035 [4] covers A records and RFC 3596 [5] covers AAAA records. An overview of DNS terminology is given in RFC 7719 [15].

3.5 Connection Establishment

As we know, the transport connection is either over Transmission Control Protocol (TCP) [16] or over Stream Control Transmission Protocol (SCTP) [17], with or without

Transport Layer Security [18, 19]. The transport connection between two peers is transparent to the "Diameter layer" and applications as soon as it gets established. There has been a notable change in the transport connection handling from the original Diameter base protocol, RFC 3588, to RFC 6733. In RFC 3588 the secure versions of the TLS/TCP or TLS/SCTP [20] transport protocols used the same port number, 3868, as their plaintext counterparts. RFC 6733 specifies that TLS/TCP and DTLS/SCTP connections must use port number 5868.[3] However, the default port numbers can be overridden by dynamic peer discovery (see Section 3.4.2).

The change of transport security handling has impacts on the transport connection implementation in general (see Chapter 5 for a detailed discussion on the security impacts), since security is not negotiated *after* the transport connection has been set up. This simplification follows the current trend of doing transport-layer security. The other notable enhancement is that the Diameter capabilities exchange messages Capability Exchange Request (CER) and Capability Exchange Answer (CEA) are no longer sent in clear text. For backward compatibility, a Diameter node may still implement the RFC 3588 way of setting up the transport-level security after the capabilities exchange.

3.5.1 The Election Process: Handling Simultaneous Connection Attempts

Since Diameter is a peer-to-peer protocol where either end of the communication can be a server or a client or both, either end can initiate peer connections. The client or server roles of a Diameter node are defined at the application level (or also at the implementation level, as a Diameter node may be developed in a way that it serves only a specific role, such as a server). It is possible that both Diameter peers initiate a peer connection establishment procedure (almost) simultaneously. For those cases, the Diameter base protocol describes an election procedure to decide which one of the two connections remains as the active peer connection. The election takes place at the transport connection receiver side during the capability exchange. The receiver of a CER message compares the DiameterIdentity contained in the `Origin-Host` AVP included in the CER message to its own DiameterIdentity. The comparison is done at the octet level comparing the two identities as blobs of octets. The peer whose DiameterIdentity is "greater" is the winner of the election. The election winner closes the transport connection that it initiated. After the election, there is only one transport connection left, which will become the peer connection between the two Diameter nodes.

3.6 Capabilities Exchange

The *capabilities exchange* occurs after the transport connection with the peer has been established. The initiator of the transport connection sends a CER message immediately to its peer after the connection establishment. The capabilities exchange has an important role in the Diameter base protocol. It is used by the adjacent peers to discover each other's identity, capabilities, configuration information, possibly supported security mechanisms/cipher-suites (for the so-called in-band security for backwards

---

3 Note that RFC 6733 erroneously uses port number 5658 everywhere except in the IANA considerations section.

compatibility), and the common set of supported application identifiers. If peers do not have a single application (or possibly a security mechanism) in common, then the responder includes an error in its CEA message (such as DIAMETER_NO_COMMON_ APPLICATION or DIAMETER_NO_COMMON_SECURITY). After sending an error-indicating CEA, the Diameter node should also close the transport connection.

Once the peers have reached the "open" state (see Section 3.8.2), the capabilities exchange usually does not happen again. Any CER message received in the open state is discarded. If a peer receives a new transport connection attempt and the subsequent CER with the same DiameterIdentity (as learned from the Origin-Host AVP) for which the peer already has a peer table entry, the peer typically rejects the CER and closes the incoming transport connection. However, there are exceptions to this. The first is *multiple connection instances*, which is discussed in detail in Section 3.9.4. Another exception is the RFC 6737 [21] extension to the base protocol that allows new applications to define whether the capabilities exchange can be rerun in order to update peer information without tearing down and then re-establishing the transport connection.

If the CER is received from an unknown peer, i.e., the DiameterIdentity is not recognized, the receiving peer may either silently discard the CER, which will eventually cause the connection attempt to time out, or respond with a CEA containing the DIAMETER_UNKNOWN_PEER error code. In both cases the receiving peer closes the transport connection. However, such stringent policy does not work well in larger Diameter deployments since every Diameter agent and server would need to know in advance all possible peers that can contact them, unnecessarily complicating the administration of such a deployment. If the Diameter deployment has a proper public key infrastructure in place, there is not a need to know peer identities in advance. The legitimacy of the other peer can be verified using cryptographic means.

3.6.1 `freeDiameter` example

Before going through the details of the messages used for the management of peer connections, here is a short hands-on example using `freeDiameter` to establish such a connection. This example assumes that you have already followed the instructions in Appendix A.

Run the following commands on the client machine to configure and start `freeDiameter`:

```
$ nw_configure.sh client.example.net
$ fD_configure.sh 3_cli
$ freeDiameterd
```

Run the following commands on the server machine to configure and start `freeDiameter`:

```
$ nw_configure.sh server.example.net
$ fD_configure.sh 3_srv
$ freeDiameterd
```

Note that this configuration is very similar to the one in Appendix A, the only difference being that the messages sent and received are displayed in a more readable form. We will use the output of those commands in the following paragraphs, which discuss the role of those commands in detail.

If we look at our `freeDiameter` log from the client.example.net vitual machine (VM), we can see the following message exchange (some lines stripped for readability):

```
SND to 'server.example.net':
    'Capabilities-Exchange-Request'
    Version: 0x01
    Length: 152
    Flags: 0x80 (R---)
    Command Code: 257
    ApplicationId: 0
    Hop-by-Hop Identifier: 0x15DAAB97
    End-to-End Identifier: 0xEA52368E
     AVP: 'Origin-Host'(264) l=26 f=-M val="client.example.net"
     AVP: 'Origin-Realm'(296) l=19 f=-M val="example.net"
     AVP: 'Origin-State-Id'(278) l=12 f=-M val=1419697829 (0x549edea5)
     AVP: 'Host-IP-Address'(257) l=14 f=-M val=192.168.35.5
     AVP: 'Vendor-Id'(266) l=12 f=-M val=0 (0x0)
     AVP: 'Product-Name'(269) l=20 f=--- val="freeDiameter"
     AVP: 'Firmware-Revision'(267) l=12 f=--- val=10200 (0x27d8)
     AVP: 'Auth-Application-Id'(258) l=12 f=-M val=4294967295 (0xffffffff
RCV from 'server.example.net':
    'Capabilities-Exchange-Answer'
    Version: 0x01
    Length: 164
    Flags: 0x00 (----)
    Command Code: 257
    ApplicationId: 0
    Hop-by-Hop Identifier: 0x15DAAB97
    End-to-End Identifier: 0xEA52368E
     AVP: 'Result-Code'(268) l=12 f=-M val='DIAMETER_SUCCESS' (2001)
     AVP: 'Origin-Host'(264) l=26 f=-M val="server.example.net"
     AVP: 'Origin-Realm'(296) l=19 f=-M val="example.net"
     AVP: 'Origin-State-Id'(278) l=12 f=-M val=1419697817 (0x549ede99)
     AVP: 'Host-IP-Address'(257) l=14 f=-M val=192.168.35.10
     AVP: 'Vendor-Id'(266) l=12 f=-M val=0 (0x0)
     AVP: 'Product-Name'(269) l=20 f=--- val="freeDiameter"
     AVP: 'Firmware-Revision'(267) l=12 f=--- val=10200 (0x27d8)
     AVP: 'Auth-Application-Id'(258) l=12 f=-M val=4294967295 (0xffffffff
```

3.6.2 The Capabilities Exchange Request

Figure 3.8 shows the CER message Command Code Format (CCF). The peer receiving the CER message learns the transport connection initiator's DiameterIdentity from the `Origin-Host` AVP and the realm from the `Origin-Realm` AVP. Note that the DiameterIdentity already contains the realm part, so the `Origin-Realm` AVP is duplicate information. The receiving peer uses the DiameterIdentity to look up the peer table for a possible existing entry.

The receiving peer looks for a mutually supported application by examining both the `Auth-Application-Id` and `Vendor-Specific-Application-Id` AVPs in the CER message. Note that the *relay Application-Id* is advertised only if the peer is a relay agent. Advertising other Application-Ids along with the relay Application-Id makes little sense. The Diameter base protocol does not explicitly prohibit such behavior, though.

```
<CER> ::= < Diameter Header: 257, REQ >
              { Origin-Host }
              { Origin-Realm }
          1* { Host-IP-Address }
              { Vendor-Id }
              { Product-Name }
              [ Origin-State-Id ]
           *  [ Supported-Vendor-Id ]
           *  [ Auth-Application-Id ]
           *  [ Inband-Security-Id ]
           *  [ Acct-Application-Id ]
           *  [ Vendor-Specific-Application-Id ]
              [ Firmware-Revision ]
           *  [ AVP ]
```

Figure 3.8 CER message format.

Another notable detail in the CER message is the mandatory presence of the Host-IP-Address AVP(s). The use case for multiple IP addresses is clear in a case of SCTP transport protocol, since the initial SCTP association does not necessarily contain all possible IP addresses the SCTP association can use during its lifetime [22] in a multi-homed/-addressed host. Including multiple IP addresses in the CER message in a case of TCP transport does not have a known "standard" use case. The situation might change with the introduction of the Multipath TCP (MPTCP) [23]. However, the use of MPTCP has not been specified for Diameter base protocol. For now, including multiple IP addresses when using the TCP transport or using MPTCP is not recommended.

Finally, since RFC 6733 deprecated the use of the Inband-Security-Id AVP for negotiating TLS, its use should be avoided unless the network administrator knows the Diameter deployment uses a mixture of RFC 6733 and RFC 3588 nodes.

3.6.3 Capabilities Exchange Answer

Figure 3.9 shows the CEA message CCF. The receiving peer responds to the CER with a CEA if it allows the transport connection to be established (the receiving peer can always reject the incoming transport connection establishment attempt). The CEA will also contain the list of application identifiers that the peer supports. Subject to the local policy both Auth-Application-Id and Vendor-Specific-Application-Id AVPs may contain either all Application-Ids supported locally, or only those that are mutually supported by both peers (with the exception of the relay Application-Id).

The rest of the CER and CEA AVPs are more or less *for further information* about the peer, and can be used for diagnostic purposes or to build a product-specific or a vendor-specific handling of the peer. It is worth mentioning that Origin-State-Id AVP can be used to determine whether the peer has recently restarted. The AVP itself is a monotonically increasing value (a counter or something derived from time) that is updated every time the Diameter node is started. If the Origin-State-Id AVP differs from a previously seen value, then the peer knows something has happened

```
<CEA> ::= < Diameter Header: 257 >
            { Origin-Host }
            { Origin-Realm }
       1* { Host-IP-Address }
            { Vendor-Id }
            { Product-Name }
            [ Origin-State-Id ]
            [ Error-Message ]
            [ Failed-AVP ]
        * [ Supported-Vendor-Id ]
        * [ Auth-Application-Id ]
        * [ Inband-Security-Id ]
        * [ Acct-Application-Id ]
        * [ Vendor-Specific-Application-Id ]
            [ Firmware-Revision ]
        * [ AVP ]
```

Figure 3.9 CEA message format.

to the other peer and that "non-Diameter state knowledge", for example of previously authenticated users, may have been lost.

3.6.4 Hop-by-Hop Identifiers

Hop-by-Hop Identifiers are used between adjacent Diameter nodes to map answer messages to request messages. The lifetime of a Hop-by-Hop Identifier is the duration of the *transaction state*. The state is removed after the node has seen an answer for a request that it has sent or forwarded earlier.

The identifiers are locally unique to a Diameter node and are always selected or generated by the Diameter node that originated or forwarded the request message. This also implies that the uniqueness of the Hop-by-Hop Identifier is tied to the peer host identity and more precisely to the peer connection identified by the host identity (the identity that also goes into the Origin-Name AVP in the CER message).

However, the Diameter node that generated the Hop-by-Hop Identifier must be prepared to receive an answer message through a different transport connection from which the request was sent.[4] There are two cases here: multiple transport connection instances between peers (see Section 3.9.4) or a collection of Diameter nodes (agents) that are able to share the hop-by-hop and transaction state information (and re-route answer messages between each other) in an implementation-dependent manner. The latter way of functioning has been known to cause interoperability issues.

Typically, if the transport connection fails in an answer message direction, it is better to discard the answer message and let the downstream nodes handle the failover procedures entirely. RFC 6733 only addresses *failover and failback procedures* to the upstream direction, i.e., from the request message failover point of view.

---

4 freeDiameter does not support this.

3.7   The Peer Table

A peer table is a data structure internal to a Diameter node that holds the information of adjacent Diameter peers with which the Diameter node has transport-level connections established. A peer table entry contains the following information:

Host Identity
Identifies the remote peer. This identity is the DiameterIdentity found in the `Origin-Host` AVP from the CER or CEA during the capability exchange. It is assumed that the host identity does not change during the lifetime of the transport connection.

Status
Provides the current state of the connection with the peer. Details on the specific states are covered in Section 3.8.2.

Static or Dynamic
Notes whether the peer table entry was dynamically discovered or manually configured.

Expiration Time
Specifies the time when the peer table entry has to be discarded or refreshed. In the case of a dynamically discovered peer table entry, the time-to-live from the DNS Resource Record (RR) specifies the time. If secure transport and public key certificates are in use, then the expiration time must not be greater than that of the certificates.

TLS/TCP and DTLS/SCTP Enabled
Indicates whether secure versions of the transport protocols are to be used when sending Diameter messages.

Security Information (optional)
Contains keys, certificates, and such as needed by the secure transport connection.

Implementation-specific Data (optional)
Contains any arbitrary information that is not covered by any Diameter specification. A good example is peer ranking information that the Diameter node uses for an educated peer selection. Another example could be storing all the IP addresses to which the remote peer DiameterIdentity resolves to.

The number of peer tables is determined by the number of peers the node is connected to or is attempting to connect to. There is a separate *peer state machine* for each peer table entry that tracks the state of the connection. The peer state machine is covered in detail in Section 3.8.2. Typically there is a single peer table entry and a transport-level connection for each remote peer. However, there are exceptions to this rule that will be discussed in more detail in Section 3.9.4.

The peer table is referenced from the *routing table*, and the intended next-hop peer node's host identity is used as the lookup key. The next-hop peer node identity is named as the *server identifier* in the routing table. The routing table is covered in more detail in Chapter 4.

The Diameter base protocol is silent on how the peer table lookup is done when multiple entries exist for the same host identity, which is the case when multiple transport connection instances are supported for the same remote peer (see Section 3.9.4 for details). At a minimum the Diameter node should able to distinguish between the transport connections based on the transport protocol level information. The local source port must be different (unless different source IP addresses are used) if the destination port and the IP address(es) are the same.

How a routing table and a peer table keep each other in sync is left to the implementation. There are situations where one will need to modify other's tables, for instance the expiration of the table entry in either the routing table or the peer table. Another example is the permanent failure of the transport connection. Although transport connections and peer connections are dependent on each other, a disconnection at the transport layer does not always imply immediate purging of the corresponding peer table entry. A Diameter node can be quite persistent trying to re-establish transport connections with peers in the event of a networking failure (refer to Section 3.8.1 and the *Tc* timer), specifically those that have a statically configured peer table entry.

3.8   Peer Connection Maintenance

The Diameter base protocol has a set of commands that are designed to maintain the connection between peers. These commands use the Application-Id 0, which is the identifier for the Diameter base protocol:

- CER and CEA (discussed in Section 3.6) are considered to be part of the generic peer connection management in the Diameter base protocol.
- Device Watchdog Request (DWR) and Device Watchdog Answer (DWA) are used to check the health and liveness of the peer connection. A Diameter node actively monitors the condition of the transport connection between peers at two levels: at the transport protocol level using the mechanisms provided by the transport protocol itself, and at the Diameter base protocol layer using watchdog messages (DWR and DWA). The Diameter layer DWR/DWA messages allow the detection of peer failures, either because the peer is unreachable or because the peer Diameter "process" has failed.
- Disconnect Peer Request (DPR) and Disconnect Peer Answer (DPA) are used to inform the other end that the peer connection is going to be closed. The DPR also contains the mandatory `Disconnect-Cause` AVP that the disconnection-initiating peer uses to inform the other peer why the transport connection is to be closed (such as REBOOTING etc). See Section 3.8.2 for more information on the DPR and DPA message exchange.

Back to our running `freeDiameter` example from 3.6.1, you can see the following DWR/DWA exchange periodically:

```
SND to 'server.example.net':
    'Device-Watchdog-Request'
      Version: 0x01
      Length: 80
      Flags: 0x80 (R---)
      Command Code: 280
      ApplicationId: 0
      Hop-by-Hop Identifier: 0x15DAAB98
      End-to-End Identifier: 0xEA52368F
       AVP: 'Origin-Host'(264) l=26 f=-M val="client.example.net"
```

```
      AVP: 'Origin-Realm'(296) l=19 f=-M val="example.net"
      AVP: 'Origin-State-Id'(278) l=12 f=-M val=1419697829 (0x549edea5
RCV from 'server.example.net':
    'Device-Watchdog-Answer'
      Version: 0x01
      Length: 92
      Flags: 0x00 (----)
      Command Code: 280
      ApplicationId: 0
      Hop-by-Hop Identifier: 0x15DAAB98
      End-to-End Identifier: 0xEA52368F
      AVP: 'Result-Code'(268) l=12 f=-M val='DIAMETER_SUCCESS' (2001)
      AVP: 'Origin-Host'(264) l=26 f=-M val="server.example.net"
      AVP: 'Origin-Realm'(296) l=19 f=-M val="example.net"
      AVP: 'Origin-State-Id'(278) l=12 f=-M val=1419697817 (0x549ede99
```

To trigger a DPR/DPA exchange and to disconnect, press Ctrl-C in the server:

```
RCV from 'server.example.net':
    'Disconnect-Peer-Request'
      Version: 0x01
      Length: 80
      Flags: 0x80 (R---)
      Command Code: 282
      ApplicationId: 0
      Hop-by-Hop Identifier: 0x5B9A6D22
      End-to-End Identifier: 0xE994D001
      AVP: 'Origin-Host'(264) l=26 f=-M val="server.example.net"
      AVP: 'Origin-Realm'(296) l=19 f=-M val="example.net"
      AVP: 'Disconnect-Cause'(273) l=12 f=-M val='REBOOTING' (0)
Peer 'server.example.net' sent a DPR with cause: REBOOTING
'STATE_OPEN'     -> 'STATE_CLOSING'        'server.example.net'
SND to 'server.example.net':
    'Disconnect-Peer-Answer'
      Version: 0x01
      Length: 80
      Flags: 0x00 (----)
      Command Code: 282
      ApplicationId: 0
      Hop-by-Hop Identifier: 0x5B9A6D22
      End-to-End Identifier: 0xE994D001
      AVP: 'Origin-Host'(264) l=26 f=-M val="client.example.net"
      AVP: 'Origin-Realm'(296) l=19 f=-M val="example.net"
      AVP: 'Result-Code'(268) l=12 f=-M val='DIAMETER_SUCCESS' (2001))
```

The messages used for connection management (CER/CEA, DPR/DPA, and DWR/DWA) are like any other Diameter messages, except that they have certain restrictions:

- These messages cannot be forwarded. They are only meant to be used between two adjacent peers. In this case the "P" flag is also unset in the Diameter message header.
- The DWR, DPR, and CER messages must not be (re)sent to an alternate peer (if a connection for such exists) upon transport failures.
- The DWR/DWA have special periodic scheduling rules and intervals. See Section 3.8.1 for details.

3.8.1 Transport Failure, Failover, and Failback Procedures

The Diameter base protocol does not describe how to recover from a transport connection failure. Rather, it references the algorithm and procedures described in RFC 3539 [24].

It is recommended that a Diameter node maintains multiple transport connections to servers or agents. The Diameter base protocol recommends maintaining at least two transport connections to *alternate* peers per realm for failover purposes. This recommendation is often also interpreted as a recommendation to have at least two transport connections between each peer, which was not the original intent. One of the transport connections is selected as the primary connection, which is used primarily for sending requests. The other connection is, naturally, used as a secondary for failover purposes. However, a Diameter node may also load balance requests among multiple peers, in which case the role of primary and secondary transport connections is no longer obvious. It is also worth noting that a Diameter node may maintain multiple secondary connections that are either *inactive* or closed, and promoted on an as-needed basis to the *active* secondary connection.

As mentioned in Section 3.8, a Diameter node monitors its peer connection status at two levels: at the transport level and at the Diameter application level. Using Diameter base protocol messaging to detect connection failures may look redundant, since the transport protocols will notify the Diameter application if a connection terminates abnormally. However, the watchdog mechanisms at the Diameter application layer enable the detection of issues with networking paths and also application failures at the Diameter peer, which would otherwise be detected only after expiration of transport-dependent timers. Once a transport failure has been detected, the Diameter node switches to the secondary (backup) transport connections, if they exist.

Figure 3.10 illustrates the Diameter transport connection failover and failback algorithm from RFC 3539 using a finite state machine. A finite state machine is a model of computer program behavior that can be in exactly one of a finite number of states at any given time. A specific event or trigger causes the state to transition to a new state. The states in the figure are shown from the peer control block (PCB) point of view. The PCB, which can be considered to be the peer state machine or at least an integral part of it, keeps state for a transport connection. The "initial" state in the illustrated state machine is the "A" state, and the exit state in a case of transport disconnection is the "B" state. Both "A" and "B" are further explained in Figure 3.11. The failover and failback state machine is entered from the state "A" when there is at least one active peer connection promoted as a (temporary) primary connection.

The transport connection can be to either a *primary* or a *secondary* peer, i.e., using primary or secondary transport connections. A Diameter node maintains a watchdog timer called Tw. The Tw timer is used to trigger timeouts at the Diameter application level. When the Tw timer expires, it triggers an attempt to re-establish a peer connection that has no existing transport connection (refer to the state machine transitions from "DOWN" to "DOWN Attempting Open" in Figure 3.11). The Tc timer, which specifies how often to attempt to make a transport connection, is actually same as the Tw timer but is used only to trigger an attempt to establish a peer connection when one does not

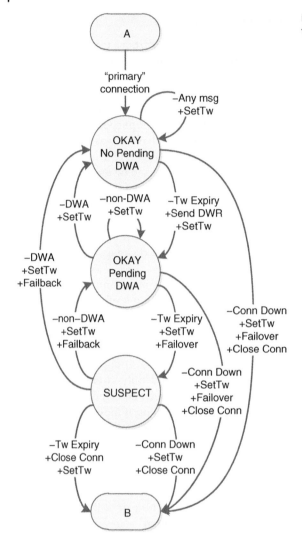

Figure 3.10 Diameter peer connection failover and failback state machine.

exist. This is shown in the state machine transition from "NO PEER CONNECTION" to "INITIAL" in Figure 3.11.

The Tw timer has an initial value *TwInit* that defaults to 30 seconds (just like the Tc timer) and must not be lower than 6 seconds. A low initial value increases the chances of duplicate messages, as well as the number of spurious failover and failback attempts. In order to avoid all PCBs in a Diameter node expiring at the same time, the formula for Tw adds intentional jitter to it. In the state machines, we refer to the formula as "SetTw":

$$Tw = TwInit - 2.0 + 4.0 * rand() \tag{3.1}$$

The Tw timer is reset any time any message from the peer is received. If the Tw timer expires, the Diameter node sends a watchdog request to the peer, if it has not already sent one, and resets the Tw timer. A Diameter node does not retransmit watchdog requests.

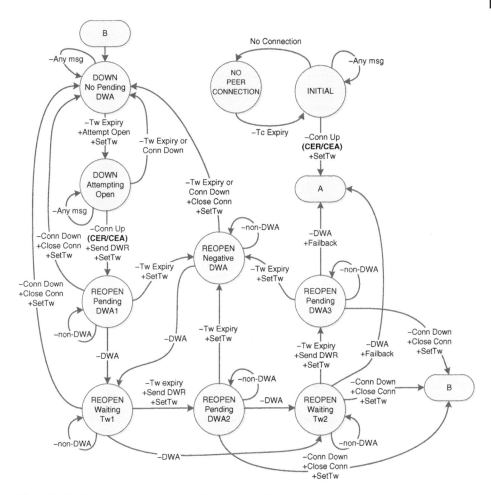

Figure 3.11 Diameter transport connection state machine.

In order for a Diameter node to failover, it must maintain a pending message queue for a given peer. When an answer message is received, the node removes the corresponding request, which can be determined by its Hop-by-Hop Identifier, from the queue. If the Tw timer expires and a watchdog response is pending, then failover is initiated. The node sends all messages in the pending message queue to an alternate peer, if one is available. Once the peer's Tw timer has expired at least twice, the node can close or reset the primary connection. The node sends subsequent requests to the alternate peer until the Tw timer on the primary connection is reset.

In the state machine, state transitions identify the *event* that triggered the transition (prefixed with "-") and the *action* that occurs during the state transition (prefixed with "+"). Table 3.1 lists the events used in the state machines.

Table 3.2 lists the actions that are taken during state transitions. It is easier to follow the state machines when you keep in mind that each PCB is equal to a transport connection and a peer connection. So, if there are two peer connections, then there are also at least

Table 3.1 State machine transition events.

Event	Description
Any msg	An arrival of any Diameter message.
Tc Expiry	An expiration of the Tc timer.
Tw Expiry	An expiration of the Tw timer.
Conn Down	The transport connection (TCP or SCTP) goes down.
DWA	An arrival of a DWA message.
non-DWA	An arrival of a Diameter message other than DWA.

Table 3.2 State machine transition actions.

Action	Description
SetTw	Reset the Tw and the timer to the initial value (see Formula 3.1).
Send DWR	Send a DWR.
Failover	Promote the secondary transport connection to the active temporary primary connection (and possibly switch a peer).
Failback	Switch back to the original primary transport connection as the active primary connection (and possibly switch a peer).
Close Conn	Close the transport connection.
Conn Up	Establish the transport connection.
Attempt Open	Initiate opening of the transport connection.

two PCBs; one being the primary and the others secondaries. Essentially, there may be multiple concurrent PCBs in the system. In the case of multiple secondary PCBs, some of them may be inactive or even closed (i.e., the actual transport protocol connection has not been established). When a *Failover* action takes place, some of the active secondary PCBs are temporarily promoted to the primary connection. When a *Failback* action takes place, the previous primary PCB is again promoted as the primary connection. Which peer connection is primary and which are secondaries is typically determined by configuration, and nothing prevents an implementation from promoting a secondary peer connection to primary status permanently at runtime.

When a failover or a failback takes place, queued requests that have not yet received a matching answer are re-sent over the transport connection that was promoted to active status. A Diameter node marks re-sent request messages with the T command flag in the message header, which gives the receiver a hint that the request message was "potentially a retransmitted message". Diameter agents that forward a request message must not change the T command flag setting.

Figure 3.11 illustrates the missing part of the Diameter transport connection state machine shown in Figure 3.10. Again, the state machine and its transitions need to be viewed from the peer control block perspective. The state machine is initially entered from the state "A" when the peer connection is first established. In the figure, the pseudo state "B" is entered only when the transport protocol connection disconnects. The pseudo state "B" is used as a "jump" within the state machine. The state machine also includes an additional state not found in RFC 3539: "NO PEER CONNECTION". This state is used only when the transport connection is established for the very first time, for example after starting up the Diameter node. The handling of failure is different for initial transport connection establishment than for a previously operational connection.

To avoid switching between the primary and secondary transport connections, the RFC 3539 algorithm tries to be rather "lazy", bringing a disconnected transport connection back to service. What is not shown in the state machine is the handling of the active and inactive/closed secondary peer connections. An implementation may silently close inactive peer connections in the background or establish new secondary connections based on internal logic. However, in order to avoid unnecessary establishing and closing transport protocol connections, RFC 3539 recommends keeping inactive connection around at least for 5 minutes.

The state machine in Figure 3.11 also shows where the CER/CEA exchanges discussed in Section 3.6 happen. Another enhancement over the state machines found in RFC 3539 are the inclusion of the Tc timer expiration and the *Conn Down* events.

3.8.2 Peer State Machine

The peer state machine is a finite state machine that an Diameter implementation uses to keep track of each peer's connection status. This peer state machine, which can be considered as the "main state machine" of the Diameter base protocol, is illustrated in Figure 3.12. It is worth noting that the peer state machine overlaps the state machines for peer connection and failover/failback to some extent (see Section 3.8.1), but the peer state machine tracks transport connections without the details of the algorithm discussed in Section 3.8.1. For example, the peer state machine connection setup follows only the initial connection establishment procedure and lacks all failover/failback details.

The peer state machine is entered from the "INITIAL" state, which corresponds to the state and transition for the initial connection setup, that is, the transition from "INITIAL" to "A" in the transport connection state machine (see Figure 3.11).

Events in the peer state machine are prefixed with "-" and are described in Table 3.3. Actions are are prefixed with "+" and are described in Table 3.4. For both the events and actions, the prefix "I-" indicates the initiator side of the connection and similarly the prefix "R-" indicates the receiver side of the connection. The notation "I/R" means that the event or action applies to both the initiator and the receiver. Note that the event *timeout* is different from the Tw timer-caused timeout, and is rather an implementation-defined value, for example to timeout a transport connection creation. Both timers may be the same, though, depending on the implementation.

3.9 Advanced Transport and Peer Topics

This section covers advanced topics on transport connection handling that may be hard to interpret unambiguously from the Diameter base protocol.

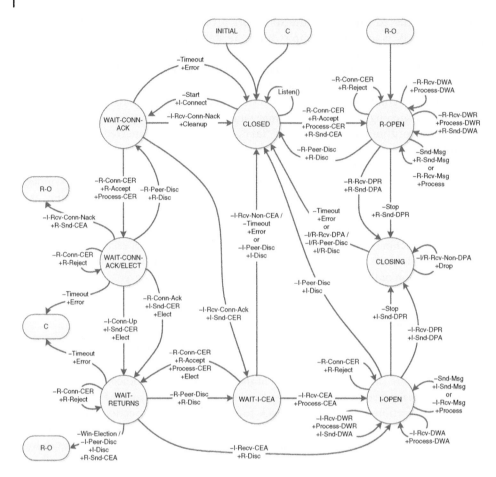

Figure 3.12 Diameter peer state machine. The pseudo states 'C' and 'R-O' are 'jumps' to the states 'CLOSED' and 'R-OPEN' respectively.

3.9.1 TCP Multi-homing

The Diameter base protocol supports multi-homing of Diameter nodes. In the Diameter context and specifically in the DiameterIdentity context that means a single DiameterIdentity may resolve in DNS to multiple IP addresses. These IP addresses may be configured into a single or multiple interfaces in a host that runs the Diameter node. If multiple IP addresses are configured and provisioned for a single DiameterIdentity, then the Diameter node must be able to originate and accept peer transport connections on all of those addresses. Note that multi-homing in this context is different from *multiple transport connection*, which is discussed in Section 3.9.4.

Interestingly, during the capability exchange (see Section 3.6) each peer informs the other about its IP addresses. Pedantically speaking, the Diameter peer has to include at least one IP address in the Host-IP-Address AVP. The Diameter base protocol does not define how these Host-IP-Address AVPs should be used for peer connection establishment. For example, the peer table (see Section 3.7) does not mandate storing

Table 3.3 Peer state machine transition events.

Event	Description
I/R-Conn-CER	A transport connection was established and a CER message received over the transport connection.
I/R-Peer-Disc	A peer connection was disconnected and an appropriate indication was received (see Section 3.8.1).
I/R-Rcv-DWR	A DWR message is received (see Section 3.8.1).
I/R-Rcv-DWA	A DWA message is received (see Section 3.8.1).
I/R-Rcv-DPR	A DPR message is received. The other peer indicates the transport connection is going to be closed.
I/R-Rcv-DPA	A DPA message is received. The other peer acknowledges it received the DPR and is ready for the transport connection to be closed.
I/R-Rcv-Disc	An indication was received that the transport connection was closed/disconnected.
I-Rcv-Conn-Ack	An acknowledgement was received that the transport connection was established.
I-Rcv-Conn-Nack	The transport connection establishment failed.
Timeout	A Diameter application-defined timer expired while waiting for some event to happen.
Start	A Diameter application has signaled internally that a peer connection needs to be established with the remote peer.
Stop	A Diameter application has signaled that a peer connection needs to be closed.
Win-Election	An election took place and the local node was the winner (see Section 3.8.2).
Snd-Msg	A node sent any application message.
I/R-Rcv-Msg	A node received any application message (i.e., not CER/CEA, DWR/DWA or DPR/DPA).
I-Rcv-Non-CEA	A message other than CEA is received.
I/R-Rcv-Non-DPA	A message other than DPA is received.

IP address information, just the remote peer DiameterIdentity. Specifically, in a case of TCP there seems to be no use for informing the other peer of anything but the IP address. Perhaps some failover solutions could use additional IP addresses for faster failover, i.e., basically skipping the DNS query. This could make sense if all peers are statically configured. In this case peers could inform the other end that there are more addresses that can be used to reach the node. However, this is just a speculation. Our recommendation is to include all possible IP addresses that the peer is going to (or able to) use, with the knowledge it has during the initial peer connection establishment, into `Host-IP-Address` AVPs and then use that set of IP addresses during the lifetime of the peer connection.

3.9.2 SCTP Multi-homing

One of the fundamental features of SCTP is the built-in support for multi-homing for enhanced reliability. Other features of SCTP include the support of multiple transport

Table 3.4 Peer state machine transition actions.

Actions	Description
I/R-Snd-DPR	Send DPR message. The peer indicates that it wants to close the transport connection.
I/R-Snd-DPA	Send DPA message and acknowledge the forthcoming closure of the transport connection.
I/R-Snd-DWR	Send DWR message, see Section 3.8.1.
I/R-Snd-DWA	Send DWA message, see Section 3.8.1.
I/R-Snd-CER	Send CER message over the newly established transport connection, see Section 3.6.
I/R-Snd-CEA	Send CEA message over the newly accepted transport connection, see Section 3.6.
I/R-Peer-Disc	The transport connection is disconnected and local resources are freed (this does not imply the Peer Control Block is freed, see Section 3.8.1).
Process-DWR	Process and act according to the received DWR, see Section 3.8.1.
Process-DWA	Process and act according to the received DWA, see Section 3.8.1.
Process-CER	Process and act according to the received CER, see Section 3.6.
Process-CEA	Process and act according to the received CEA, see Section 3.6.
Process	Process an application-level Diameter message (i.e., not CER/CEA, DWR/DWA or DPR/DPA).
Error	The transport connection is disconnected either abruptly or in a clean way. Local resource related to the transport connection are freed (this does not imply that the Peer Control Block is freed, see Section 3.8.1).
Elect	The election takes place, see Section 3.8.2.
R-Accept	Accept an incoming transport connection (i.e., the connection that the CER arrives from).
R-Reject	Reject an incoming transport connection (i.e., the connection that the CER arrived from).
Cleanup	Local resources related to the transport connection are freed and, if needed, the transport connection is also closed. This does not imply that the Peer Control Block is freed, see Section 3.8.1).
Drop	Silently discard the received Diameter message.

streams and the out-of-order but reliable delivery of messages. These features make SCTP a superior transport protocol. RFC 6733 Section 2.1.1 details how to deal with SCTP multi-streaming and out-of-order delivery in order to avoid the head-of-line blocking issue (an issue with TCP transport) and possible out-of-order delivery concerns during the peer connection establishment that may cause a race condition and result in closing of the peer connection.

An SCTP association contains one or more IP addresses that the association can use if the primary path between endpoints fails (path here is also assumed to include the local networking interface). There is also an extension to SCTP multi-homing to allow the dynamic addition and removal of IP addresses from the SCTP association [22]. Although this extension is not part of the Diameter base protocol, nothing prohibits

its implementation, and a Diameter peer may agree on using it at the SCTP transport level, which is out of the scope of the Diameter base protocol.

How do the `Host-IP-Address` AVPs negotiated during the capability exchange (see Section 3.6) relate to SCTP and multi-homing? There is no clear use case and RFC 6733 is silent on this. One possible use could be additional checks (for "security" purposes) on whether or not incoming packets are from legitimate sources. Again, these are just speculations. Our recommendation is to include all possible IP addresses the peer is going to (or able to) use, with the knowledge it has during the initial peer connection establishment, into `Host-IP-Address` AVPs and then use that set of IP addresses during the lifetime of the peer connection.

3.9.2.1 Multi-homing in `freeDiameter`

`freeDiameter` does not support TCP multi-homing at the moment; only one transport-level connection per DiameterIdentity is allowed, which simplifies the state machine and the routing mechanism to some extent. However, SCTP connections are natively supported on operating systems that provide the low-level feature, allowing for multi-homing capability.

Here is a simple example setup showing SCTP multi-homing in operation, using two VMs interconnected with two independent networks (192.168.35.0/24 on eth0 and 192.168.65.0/24 on eth1).

1. On the first VM, issue the following commands to set up the configuration:

```
$ nw_configure.sh server.multi.example.net
$ fD_configure.sh 3_srv_multi
```

2. On the second VM, issue the following:

```
$ nw_configure.sh client.multi.example.net
$ fD_configure.sh 3_cli_multi
```

3. Start `freeDiameter` on the server VM by issuing the command `$ freeDiameterd`.

The following lines from the console output are worth noting:

```
...
Number of SCTP streams . : 10
...
Local server address(es): 192.168.35.10{---L-}  192.168.65.10{---L-}
freeDiameterd daemon initialized.
```

After starting `freeDiameter` on the client VM, the two machines establish a SCTP connection and the CER/CEA exchange is as follows.

```
RCV from '<unknown peer>':
    'Capabilities-Exchange-Request'
      Version: 0x01
      Length: 192
      Flags: 0x80 (R---)
      Command Code: 257
      ApplicationId: 0
      Hop-by-Hop Identifier: 0x72CF4EE9
      End-to-End Identifier: 0x3A9BBBE8
       AVP: 'Origin-Host'(264) l=32 f=-M val="client.multi.example.net"
```

```
      AVP: 'Origin-Realm'(296) l=25 f=-M val="multi.example.net"
      AVP: 'Origin-State-Id'(278) l=12 f=-M val=1424585641 (0x54e973a9)
      AVP: 'Host-IP-Address'(257) l=14 f=-M val=192.168.35.5
      AVP: 'Host-IP-Address'(257) l=14 f=-M val=192.168.65.5
      AVP: 'Vendor-Id'(266) l=12 f=-M val=0 (0x0)
      AVP: 'Product-Name'(269) l=20 f=--- val="freeDiameter"
      AVP: 'Firmware-Revision'(267) l=12 f=-- val=10200 (0x27d8)
      AVP: 'Inband-Security-Id'(299) l=12 f=-M val='NO_INBAND_SECURITY'
      AVP: 'Auth-Application-Id'(258) l=12 f=-M val=4294967295 (0xffffffff
SND to 'client.multi.example.net':
  'Capabilities-Exchange-Answer'
    Version: 0x01
    Length: 192
    Flags: 0x00 (----)
    Command Code: 257
    ApplicationId: 0
    Hop-by-Hop Identifier: 0x72CF4EE9
    End-to-End Identifier: 0x3A9BBBE8
      AVP: 'Result-Code'(268) l=12 f=-M val='DIAMETER_SUCCESS' (2001)
      AVP: 'Origin-Host'(264) l=32 f=-M val="server.multi.example.net"
      AVP: 'Origin-Realm'(296) l=25 f=-M val="multi.example.net"
      AVP: 'Origin-State-Id'(278) l=12 f=-M val=1424585294 (0x54e9724e)
      AVP: 'Host-IP-Address'(257) l=14 f=-M val=192.168.35.10
      AVP: 'Host-IP-Address'(257) l=14 f=-M val=192.168.65.10
      AVP: 'Vendor-Id'(266) l=12 f=-M val=0 (0x0)
      AVP: 'Product-Name'(269) l=20 f=--- val="freeDiameter"
      AVP: 'Firmware-Revision'(267) l=12 f=-- val=10200 (0x27d8)
      AVP: 'Auth-Application-Id'(258) l=12 f=-M val=4294967295 (0xffffffff
```

As expected, both peers advertise their local IP endpoints in a Host-IP-Address AVP. If we look at the detail of the SCTP communication, using, for example, Wireshark, we see the following.

1. The SCTP connection is established using the primary IP address; however, the INIT and INIT_ACK messages contain the additional IP addresses.

```
Internet Protocol Version 4, Src: 192.168.35.5, Dst: 192.168.35.10
Stream Control Transmission Protocol, Src Port: 42344, Dst Port: 3868
    INIT chunk (Outbound streams: 10, inbound streams: 65535)
        Chunk type: INIT (1)
        Number of outbound streams: 10
        Number of inbound streams: 65535
        Initial TSN: 2671975453
        IPv4 address parameter (Address: 192.168.35.5)
        IPv4 address parameter (Address: 192.168.65.5)
        Supported address types parameter (Supported types: IPv6, IPv4)

Internet Protocol Version 4, Src: 192.168.35.10, Dst: 192.168.35.5
Stream Control Transmission Protocol, Src Port: 3868, Dst Port: 42344
    INIT_ACK chunk (Outbound streams: 10, inbound streams: 10)
        Chunk type: INIT_ACK (2)
        Number of outbound streams: 10
        Number of inbound streams: 10
        Initial TSN: 1541122988
        IPv4 address parameter (Address: 192.168.35.10)
        IPv4 address parameter (Address: 192.168.65.10)
        State cookie parameter (Cookie length: 260 bytes)
```

2. After the COOKIE_ECHO / COOKIE_ACK exchange, the CER/CEA happens as displayed above.
3. Periodically, the SCTP layer exchanges HEARTBEAT messages with all the IP addresses of the association to check their reachability status.

```
Internet Protocol Version 4, Src: 192.168.65.10, Dst: 192.168.65.5
Stream Control Transmission Protocol, Src Port: 3868, Dst Port: 42344
    HEARTBEAT chunk (Information: 44 bytes)

Internet Protocol Version 4, Src: 192.168.65.5, Dst: 192.168.65.10
Stream Control Transmission Protocol, Src Port: 42344, Dst Port: 3868
    HEARTBEAT_ACK chunk (Information: 44 bytes)

Internet Protocol Version 4, Src: 192.168.35.5, Dst: 192.168.35.10
Stream Control Transmission Protocol, Src Port: 42344, Dst Port: 3868
    HEARTBEAT chunk (Information: 44 bytes)

Internet Protocol Version 4, Src: 192.168.35.10, Dst: 192.168.35.5
Stream Control Transmission Protocol, Src Port: 3868, Dst Port: 42344
    HEARTBEAT_ACK chunk (Information: 44 bytes)
```

4. As long as the primary IP address is reachable, it is used for the Diameter message payload exchanges.
5. If at some point the primary IP addresses become unreachable, the association will smoothly switch to the secondary address, provided that it is still reachable. This can be emulated easily with VirtualBox by "disconnecting" the primary network interface. Right-click the small network icon at the lower right part of the VM window and click "Connect Network Adapter 1" to break the primary link (Figure 3.13). The HEARTBEAT exchange continues only on the secondary IP and the next Diameter payload message is sent to that IP.

```
Internet Protocol Version 4, Src: 192.168.65.5, Dst: 192.168.65.10
Stream Control Transmission Protocol, Src Port: 42344, Dst Port: 3868
Diameter Protocol
    Command Code: 280 Device-Watchdog
    ApplicationId: 0
    AVP: Origin-Host(264) l=32 f=-M- val=client.multi.example.net
    AVP: Origin-Realm(296) l=25 f=-M- val=multi.example.net
    AVP: Origin-State-Id(278) l=12 f=-M- val=1424585641
```

Note that for the above example we disabled the transport security mechanism (TLS/DTLS) in the configuration file to make it easier to spy the traffic with Wireshark.

Figure 3.13 VirtualBox interface to disable a virtual network link.

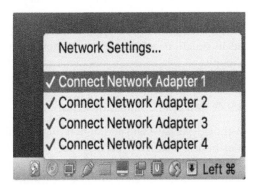

3.9.3 Avoiding Head-of-Line Blocking

Head-of-line blocking is a situation where transport protocol level retransmissions delay the delivery of all subsequent messages until the retransmitted message gets through. As mentioned in Section 3.9.2, RFC 6733 Section 2.1.1 has detailed language for handling of head-of-line blocking [24] in the Diameter message delivery. In the case of Diameter's use of TCP transport, there is no way of avoiding head-of-line blocking. It could be argued that a clever use of the *multiple connection instances* between two Diameter peers could be used to mimic the multiple transport streams feature of SCTP and work around issues caused by the head-of-line blocking that way.

In the case of SCTP, there are two ways to solve head-of-line blocking: use of multiple streams or use of the out-of-order delivery feature of the SCTP. RFC 3588 recommended using multiple streams and disabling out-of-order message delivery. Ideally, if each Diameter application between two peers used its own streams, then possible retransmissions at the transport level would not interfere with other applications, only the one experiencing lost packets. The problem with this approach is that both ends of the transport connection had to agree on the number of streams to use (the number of streams may be different in uplink and downlink directions). Unfortunately many implementations chose the easy way out by supporting only a single stream, which effectively makes SCTP behave like TCP when it comes to head-of-line blocking.

RFC 6733, on the other hand, recommends using a single stream but enabling the out-of-order message delivery. In this way messages sent after the "message in retransmission" can bypass it and thus avoid head-of-line blocking. The use of out-of-order message delivery can have a rare downside during the establishment of a peer connection. If the peer that received the CER message sends to the connection initiator an application message immediately after sending the CEA, it is possible for the application message to arrive before the CEA. Since the connection initiator is in *WAIT-I-ACK* state, the reception of the non-CEA message will cause the initiator to error and close the transport connection (see Figure 3.12 in Section 3.8.2). A similar race condition may happen when a DPR is sent immediately after an application message. Due to possible message reordering, the peer may receive the DPR before the application message, and the application message is discarded before processing (the received peer state machine being in the *CLOSING* state).

There are multiple ways of mitigating the issues caused by reordering. First, stick with ordered delivery of SCTP messages and use a single stream instead of multiple streams. Second, if out-of-order delivery is still desired, then use small delays in the above-mentioned message sequences (e.g., wait a small amount of time before sending a DPR after sending an application message). Third, switch the out-of-order delivery off in situations where there are known issues. All these can be categorized as hacks, though.

3.9.4 Multiple Connection Instances

Although we have said earlier that there is only one peer connection between two Diameter nodes, Diameter has a concept of *multiple connection instances*. This allows two Diameter peers to have any number of transport connections between them with a precondition that both ends support the feature. Furthermore, RFC 6733 says in Section 2.1:

> A given Diameter instance of the peer state machine MUST NOT use more
> than one transport connection to communicate with a given peer,
> unless multiple instances exist on the peer, in which case a
> separate connection per process is allowed.

The above text indicates multiple instances of the Diameter node *process* also exist (we could also say that multiple threads of the Diameter node exist). The text refers to the UNIX way of forking client and server processes, i.e., using *fork()* to create a new client or server instance of the node with its own peer table entry. Obviously the client side (i.e., the connection initiator) has to bind the transport connection socket to a new source port before attempting a connection to the remote peer, otherwise the connection attempt will fail. On the server side, the connection receiver has to implement the arriving connection logic so that when the new instance is created and the connection is accepted, the new peer table entry state machine starts from the *CLOSED* state instead of whatever state the Diameter node "instance" may be in at that time, otherwise according to the peer state machine (see Section 3.8.2, Figure 3.12) the incoming connection is rejected. This implies that the peer table is not searched for existing entries with the same *Host Identity* found in the arriving CER's `Origin-Host` AVP when a new transport connection is established. The key here to make multiple instances to work is that each peer transport connection pair needs to have their own instance of peer control block (PCB) with its respective peer state machines.

Figure 3.14 illustrates how TCP connections can be used to realize multiple connection instances between two peers. Typically the transport connections are between one IP address pair. However, a DiameterIdentity may resolve to multiple IP addresses and that can be used when creating multiple connection instances also. When multiple IP addresses are used, the connection initiator can reuse the same source port number for its transport connections. These are implementation details, though.

To summarize, the following has to be supported by both Diameter peers for the *multiple connection instances* to work:

- Both the transport connection initiator (client) and the receiver (server) have to support the feature.
- The connection initiator has to originate the transport connection from a different source port than any existing and established transport connection between the Diameter peers. This requirement can be relaxed if the originator DiameterIdentity resolves to multiple IP addresses and each new "multiple connection instance" connection uses a different source IP address.
- The connection initiator must create a new peer table entry whose peer state machine starts from the *CLOSED* state.
- The connection receiver has to skip the peer table lookup for an existing peer table entry for the same *Host Identity* as found in the incoming CER `Origin-Host` AVP.
- The connection receiver needs to create a new peer table entry for the new connection and start from the peer state machine *CLOSED* state.

RFC 6733 is silent on how multiple connection instances are handled from the routing and forwarding point of view. How the Diameter node implementation selects the peer connection to originate or forward messages is left for the implementation. In the case of multiple connection instances, the routing table would point to multiple peer table

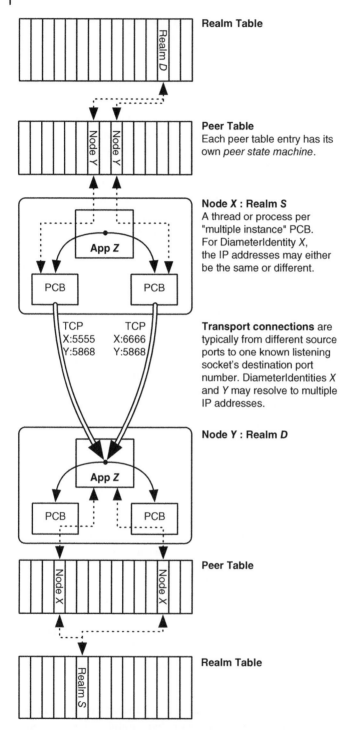

Realm Table

Peer Table
Each peer table entry has its
own *peer state machine*.

Node *X* : Realm *S*
A thread or process per
"multiple instance" PCB.
For DiameterIdentity *X*,
the IP addresses may either
be the same or different.

Transport connections are
typically from different source
ports to one known listening
socket's destination port
number. DiameterIdentities *X*
and *Y* may resolve to multiple
IP addresses.

Node *Y* : Realm *D*

Peer Table

Realm Table

Figure 3.14 Relations between realm tables, peer tables, application, and peer control blocks in a case
of multiple connection instances.

entries and the peer table would have multiple entries with the same remote peer *Host Identity*.

Multiple connection instances are unlikely to improve transport level reliability, because each instance exchanges IP packets between the same source and destination IP pair, thus using the same network path. However, multiple instances could be used for host internal load balancing, assuming the new instances are instantiated in a new CPU core or on a blade in a cluster, or similar. Since the multiple connection instances behavior is somewhat underspecified in RFC 6733, it is likely that proper "out of box" interoperability would be challenging to achieve between Diameter implementations.

References

1 J. Veizades, E. Guttman, C. Perkins, and S. Kaplan. Service Location Protocol. RFC 2165, Internet Engineering Task Force, June 1997.

2 M. Mealling and R. Daniel. The Naming Authority Pointer (NAPTR) DNS Resource Record. RFC 2915, Internet Engineering Task Force, Sept. 2000.

3 A. Gulbrandsen, P. Vixie, and L. Esibov. A DNS RR for specifying the location of services (DNS SRV). RFC 2782, Internet Engineering Task Force, Feb. 2000.

4 P. Mockapetris. Domain names – implementation and specification. RFC 1035, Internet Engineering Task Force, Nov. 1987.

5 S. Thomson, C. Huitema, V. Ksinant, and M. Souissi. DNS Extensions to Support IP Version 6. RFC 3596, Internet Engineering Task Force, Oct. 2003.

6 M. Jones, J. Korhonen, and L. Morand. Diameter Straightforward-Naming Authority Pointer (S-NAPTR) Usage. RFC 6408, Internet Engineering Task Force, Nov. 2011.

7 M. Mealling. Dynamic Delegation Discovery System (DDDS) Part One: The Comprehensive DDDS. RFC 3401, Internet Engineering Task Force, Oct. 2002.

8 L. Daigle and A. Newton. Domain-Based Application Service Location Using SRV RRs and the Dynamic Delegation Discovery Service (DDDS). RFC 3958, Internet Engineering Task Force, Jan. 2005.

9 J. Korhonen, H. Tschofenig, J. Bournelle, G. Giaretta, and M. Nakhjiri. Diameter Mobile IPv6: Support for Home Agent to Diameter Server Interaction. RFC 5778, Internet Engineering Task Force, Feb. 2010.

10 M. Cotton, L. Eggert, J. Touch, M. Westerlund, and S. Cheshire. Internet Assigned Numbers Authority (IANA) Procedures for the Management of the Service Name and Transport Protocol Port Number Registry. RFC 6335, Internet Engineering Task Force, Aug. 2011.

11 T. Berners-Lee, R. Fielding, and L. Masinter. Uniform Resource Identifier (URI): Generic Syntax. RFC 3986, Internet Engineering Task Force, Jan. 2005.

12 D. Crocker and P. Overell. Augmented BNF for Syntax Specifications: ABNF. RFC 2234, Internet Engineering Task Force, Nov. 1997.

13 C. Rigney, S. Willens, A. Rubens, and W. Simpson. Remote Authentication Dial In User Service (RADIUS). RFC 2865, Internet Engineering Task Force, June 2000.

14 C. Finseth. An Access Control Protocol, Sometimes Called TACACS. RFC 1492, Internet Engineering Task Force, July 1993.

15 P. Hoffman, A. Sullivan, and K. Fujiwara. DNS Terminology. RFC 7719, Internet Engineering Task Force, Dec. 2015.

16 J. Postel. Transmission Control Protocol. RFC 0793, Internet Engineering Task Force, Sept. 1981.

17 R. Stewart. Stream Control Transmission Protocol. RFC 4960, Internet Engineering Task Force, Sept. 2007.

18 T. Dierks and E. Rescorla. The Transport Layer Security (TLS) Protocol Version 1.2. RFC 5246, Internet Engineering Task Force, Aug. 2008.

19 E. Rescorla and N. Modadugu. Datagram Transport Layer Security Version 1.2. RFC 6347, Internet Engineering Task Force, Jan. 2012.

20 A. Jungmaier, E. Rescorla, and M. Tuexen. Transport Layer Security over Stream Control Transmission Protocol. RFC 3436, Internet Engineering Task Force, Dec. 2002.

21 K. Jiao and G. Zorn. The Diameter Capabilities Update Application. RFC 6737, Internet Engineering Task Force, Oct. 2012.

22 R. Stewart, Q. Xie, M. Tuexen, S. Maruyama, and M. Kozuka. Stream Control Transmission Protocol (SCTP) Dynamic Address Reconfiguration. RFC 5061, Internet Engineering Task Force, Sept. 2007.

23 A. Ford, C. Raiciu, M. Handley, and O. Bonaventure. TCP Extensions for Multipath Operation with Multiple Addresses. RFC 6824, Internet Engineering Task Force, Jan. 2013.

24 B. Aboba and J. Wood. Authentication, Authorization and Accounting (AAA) Transport Profile. RFC 3539, Internet Engineering Task Force, June 2003.

4

Diameter End-to-End Communication

4.1 Introduction

We discussed the communication between neighboring peers in Chapter 3, and specifically Section 3.2 showed basic concepts of the Diameter session spanning multiple Diameter nodes. This chapter looks at the details of communication between two Diameter nodes that are not adjacent.

4.2 The Routing Table

The *routing table* is a data structure internal to the Diameter node that contains information on how to handle Diameter request messages: either consuming the request message locally or processing the request further before routing it to the appropriate adjacent peer. There is typically one routing table per Diameter node; however, multiple routing tables are possible when routing based on policy configuration.

Each routing table entry points to one or more peer table entries. Typically there is a single peer table entry for each routing table entry. However, as discussed in Chapter 3 a single destination may have multiple peer control blocks (PCB), for example in the case of multiple connection instances. How the Diameter application and the routing table lookup selects the peer table entry in this case is implementation specific. Aspects such as peer load information or other priorities may affect the selection.

Different types of Diameter nodes use the routing table differently. Diameter clients use the routing table only to find an adjacent peer to whom to forward the originated request message. Diameter servers consume the received request message locally and do not forward them further. Diameter agents carry out additional processing to received requests before routing them to the appropriate peer.

Although some of the routing table contents are implementation specific, the following elements *must* be found in each routing table entry:

Realm Name
The realm name is the primary key for the adjacent peer lookups. The realm may be matched against the lookup key realm using exact or "longest match from the right". Depending on the implementation, the lookup key can be more complex with additional context-specific information. There may be a default, wildcarded

Diameter: New Generation AAA Protocol – Design, Practice, and Applications, First Edition.
Hannes Tschofenig, Sébastien Decugis, Jean Mahoney and Jouni Korhonen.
© 2019 John Wiley & Sons Ltd. Published 2019 by John Wiley & Sons Ltd.

entry that matches everything. The utilization of wildcards and default entries is implementation specific.

Application-Id(s)

The secondary lookup key for the adjacent peer lookups. If the next hop is a relay agent, the value of the Application-Id can be considered as a wildcard matching all applications. The implementation of wildcards and default entries is implementation specific.

Server identifier

The server identifier is the link/index to one or more peer table entries, where it is present in the peer table as the *Host Identity* field. When the *Local Action* (see below) is set to RELAY or PROXY, this field contains the DiameterIdentity of the server(s) to which the request message must be routed. When the *Local Action* field is set to REDIRECT, this field contains the identity of one or more servers to which the request message must be redirected.

Static or Dynamic

This indicates whether the entry was statically configured or was created as a result of dynamic peer discovery.

Expiration Time

Specifies when the entry expires. For example, for DNS-based dynamic peer discovery, the discovered peer information has an associated lifetime from the DNS response. If the transport security utilizes public key certificates, then the *Expiration Time* must not be greater than the lifetime of the associated certificates.

Local Action

Indicates that the node should take one of the following actions:

LOCAL

The node should consume and process the message locally. The request message has reached its final destination. That is, the realm in the routing table matches the destination realm and specifically the destination host in the request message.

PROXY

The node should route the request message to the appropriate adjacent peer node as indicated by the *Server Identifier*. A node acting as a proxy may also apply local processing to the request message based on the local policy and configuration. This processing may involve modifying the request message by adding, removing or modifying its AVPs. However, the proxy must not reorder AVPs.

RELAY

The node should route or forward the request message to the appropriate peer node as indicated by the *Server Identifier*. A node acting as a relay must not modify or reorder the request message AVPs other than to update *routing AVPs*. See Section 4.3.1 for more information.

REDIRECT

The node should give the request message the *redirecting* treatment, by sending an "error" answer message to the message originator that contains one or more destination hosts where the request can be resent and guidance on how to treat subsequent request messages to the same original destination.

Although the routing table has *actions* that could also apply to answer messages (such as PROXY), RFC 6733 does not give guidance on how to process and, specifically, "proxy process" answer messages. The *transaction state* maintained for the answer messages is separate from the routing table and possible "proxy process" modifications to the answer messages are left for the implementations to solve.

4.3 Diameter Request Routing

Routing decisions are made based on the Diameter Application-Id and the destination realm (and possibly by a number of vendor- and implementation-specific methods – there will be examples later). The lookup for the next-hop peer goes through the routing table. The procedure by which a node sends the request message to a specific peer using the destination host information found in the peer table rather than the destination realm is called *request forwarding*. The small distinction in the routing and forwarding procedures implies that the Diameter node has to consult both peer table and routing table to determine the correct treatment for the request message.

When a Diameter node routes (or forwards) a request message, it has to maintain the transaction state and buffer the message until the corresponding answer message has arrived or the transaction times out entirely.

The AVPs used for routing (and forwarding) Diameter request messages are discussed in detail in Section 4.3.1. Although Diameter implementations may use more intelligent approaches to select their next-hop peers when routing or forwarding request messages, the following simple rules regarding the AVPs in the context of routing, proxying, and forwarding should be kept in mind:

1. If the request message is to be proxied, it should contain the `Destination-Realm` and/or the `Destination-Host` AVPs. Both AVPs contain realm information, which can enable the routing of the request message. However, RFC 6733 is not clear whether request messages should contain only the `Destination-Host` AVP. Our interpretation is that such request messages cannot be proxied but are meant for direct adjacent peer communication for non-base Diameter applications. The description of the DIAMETER_UNABLE_TO_DELIVER error status code also supports this interpretation (see Section 4.4).
2. If the request message must be sent to a specific Diameter node in a specific realm, then the request message must contain both the `Destination-Realm` and the `Destination-Host` AVPs.
3. If the request message has to reach a specific realm but not a specific Diameter node, then the request message must contain only the `Destination-Realm` AVP.
4. If the request message is only for the adjacent peer and is of the base protocol "common application", such as CER or DWR, it must not contain either the `Destination-Realm` or the `Destination-Host` AVPs.
5. In addition to the standard request-routing principles and AVPs, there are specifications and implementations that also look at other AVPs when making routing and forwarding decisions. For instance, typically the `User-Name` AVP is used for the next-hop peer selection, but it can be used to populate or manipulate the `Destination-Realm` AVP. A good example can be found in mobile networks, where an International Mobile Subscriber Identity (IMSI) may both identify the subscriber and affect the selection of the next-hop peer [1]. See Section 4.3.1 for more information.

It is also possible for intermediate agents to manipulate or add AVPs to the request messages so that a desired destination gets reached. This may be done for load balancing purposes. For example, a proxy agent that load balances in a multi-level agent hierarchy could add the `Destination-Host` AVP to request messages that only contain the `Destination-Realm` AVP to ensure that the request messages reach a specific host in that realm.

4.3.1 AVPs to Route Request Messages

The Diameter request message routing relies on both the destination realm and the Application-Id. While reaching the final destination, the destination host, if present, is used to forward the message to the correct peer node.

4.3.1.1 `Destination-Realm` AVP

The `Destination-Realm` AVP is of type DiameterIdentity and contains the domain portion (i.e., the realm) of the intended request destination. When querying the routing table, the `Destination-Realm` AVP is used as one of the lookup keys along with the Application-Id. The `Destination-Realm` AVP is included only in the request messages, never in the answer messages. The `Destination-Realm` AVP has to be in the request message if proxying or relaying the message is desired. This implies that request messages that are only meant to be used between two adjacent peers (such as CER/CEA, etc.) must not have the `Destination-Realm` AVP.

The `Destination-Realm` AVP in a request message usually should not be modified as it travels to the final receiver. However, this rule has been relaxed by RFC 5729 [2] in order to accommodate realm granularity source routing of request messages using *NAI Decoration*. See Section 4.3.1 for further details.

4.3.1.2 `Destination-Host` AVP

The `Destination-Host` AVP is type of DiameterIdentity and contains the fully qualified domain name (FQDN) of the Diameter node. Thus the `Destination-Host` AVP contains both the specific Diameter node identity within a realm and the realm itself. If both the `Destination-Realm` AVP and the `Destination-Host` AVP are present in a request message, they must not contain conflicting information. Intuitively, routing and proxying the request message should be possible using the `Destination-Host` AVP alone. However, this is not the case. The `Destination-Host` AVP is used solely to name a specific Diameter node within a realm identified by the `Destination-Realm` AVP.

Similar to the `Destination-Realm` AVP, the `Destination-Host` AVP is meant to be present only in request messages that can be proxied.

4.3.1.3 `Auth-Application-Id` and `Acct-Application-Id` AVPs

Both the `Auth-Application-Id` and the `Acct-Application-Id` AVPs are of type Unsigned32 and contain the numerical identifier value of the Diameter application. If either the `Auth-Application-Id` or the `Acct-Application-Id` AVPs are present in a Diameter message other than CER and CEA, the value of these AVPs must match the Application-Id present in the Diameter message header, but they are redundant. They are listed in RFC 6733 for backward compatibility purposes but serve no purpose since the Diameter message header already contains the Application-Id information. One could argue that having the Application-Id at the message level provides cleaner layering between the application and peer connection logic.

Application-Ids are used as the other lookup key along with the `Destination-Realm` AVP into the routing table.

4.3.1.4 `User-Name` AVP

Unlike RADIUS [3], Diameter does not rely on the `User-Name` AVP for request routing purposes. However, a Diameter node may use the *User-Name* value to determine the destination realm. The `User-Name` AVP is of type UTF8String and contains the subscriber username in a format of Network Access Identifier (NAI) [4, 5], which is constructed like an email address. The NAI allows *decoration*, that is, one can embed a source route in a form of realms into the NAI user-name portion. See Figure 4.1 for examples of NAI decoration. RFC 5729 [2] updates the Diameter base protocol to explicitly support source routing based on NAI decoration. However, deployment experience has shown that NAI decoration is not a scalable and maintainable solution in larger multi-vendor and multi-operator deployments. Populating and maintaining client devices with exact AAA routing information is burdensome, and repairing breakage due to stale source routes is slow since client devices, rather than the routing infrastructure, need to be updated with new routes. Realm-level redirections [6] or a dynamic DNS-based discovery [7, 8] may be used to circumvent *stale realms* in the routing network, but there is not much deployment experience using these techniques in the case of source routing of messages.

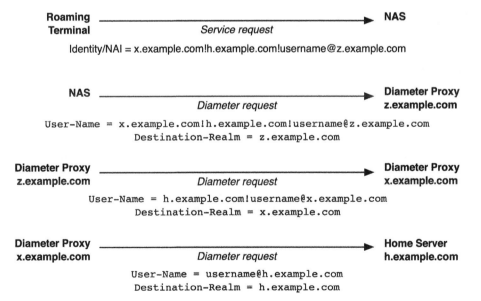

Figure 4.1 The roaming terminal decides that the Diameter messages must be routed via *z.example.com* and *x.example.com* to *h.example.com*. Example taken from RFC 5729.

4.3.2 Routing AVPs

Diameter end-to-end communication relies on a number of *routing AVPs*. Unlike what the category name suggests, these AVPs are not used for the request routing or forwarding, but to record the used route and the state information of the traversed agents.

Note that even relay agents process the routing AVPs. Both relay and proxy agents append the Route-Record AVP to the request. The AVP contains the DiameterIdentity of the agent. Another point to stress here is that the Route-Record AVPs are intended to be included only in the request message, whereas the received Proxy-Info AVPs are echoed in the answer message in the same order from the corresponding request message.

Routing AVPs serve three main purposes:

1. Relay and proxy agents inspect the array of Route-Record AVPs in the request message in order to detect routing loops.
2. The receiver of the request message can carry out *path authorization* by verifying that the request message traversed through realms and nodes that are supposed to be on the path. Note the lack of cryptographic origin authentication of the inserted Route-Record AVPs – it is trivial to spoof or modify the routing AVPs if some intermediate wishes to do. This "feature" is commonly used, for example to realize topology hiding.
3. The Proxy-Info AVPs serve the purpose of recording and remembering the *transaction state* of the traversed *stateless* agents.

4.3.2.1 Route-Record AVP
The Route-Record AVP is of type DiameterIdentity and contains the identity of the Diameter node (i.e., the agent) that inserted the AVP. The Route-Record AVP is used only in request messages, and its content must be the same as the Diameter node's Origin-Host AVP that is used in the CER message during the capability exchange.

4.3.2.2 Proxy-Info AVP
The Proxy-Info AVP is of type Grouped and contains two sub-AVPs: the Proxy-Host and the Proxy-State AVPs. The Proxy-Info AVP is shown in Figure 4.2. Despite what the AVP name suggests, the AVP is meant not only for proxy agents but also for relay agents.

The Proxy-Host AVP is of type DiameterIdentity and contains the identity of the Diameter node that inserted the Proxy-Info AVP. The Proxy-State AVP is of type OctetString and contains an opaque octet blob that is only "meaningful" to the node that inserted it. RFC 6733 also recommends using cryptography to protect its content.

```
Proxy-Info ::= < AVP Header: 284 >
                 { Proxy-Host }
                 { Proxy-State }
             *   [ AVP ]
```

Figure 4.2 The Grouped Proxy-Info AVP and its sub-AVPs.

The Diameter node that inserted the `Proxy-Info` AVP into the request message is also responsible for removing it from the answer message before forwarding the answer.

4.4 Request Routing Error Handling

4.4.1 Detecting Duplicated Messages

The End-to-End Identifier is used for detecting duplicated messages. The *End-to-End Identifier* is a 32-bit integer placed in the Diameter message header by the *request message originator* and is locally unique to the *Diameter node that created it*. RFC 6733 recommends creating the End-to-End Identifier out of two parts: placing the low-order bits of the current time into the upper 12 bits of the 32-bit value and using a random value for the lower 20 bits. RFC 6733 does not define "current time". If "current time" is based on Network Time Protocol (NTP), then the lower 12 bits are parts of the "fraction of second" in any of the NTP on-wire formats [9]. The End-to-End Identifier must remain unique to its creator for at least 4 minutes, even across reboots.

Intermediate Diameter agents (relays, redirects, proxies) are not allowed to modify the value. In the answer direction, the value of the End-to-End Identifier is copied to the answer message.

In addition to using a combination of the End-to-End Identifier and the `Origin-Host` AVP to detect duplicated messages, the message receiver can look for the *T flag* in the request command flag field. The message originator may set the T flag if it is retrying messages after a transport failure or after a reboot. The 4-minute minimum value for the End-to-End Identifier uniqueness hints to the Diameter server that it should be prepared to "remember" received requests for that period of time. The same also applies for the request originator regarding answer messages.

If a server receives a duplicate Diameter request message, it should reply with the same answer. This requirement does not concern transport-level identifiers and parameters such as the *Hop-by-Hop Identifier* and routing AVPs (see Section 4.3.1). Furthermore, the reception of the duplicate message should not cause any state transition changes in the peer state machine.

4.4.2 Error Codes

The Diameter base protocol has a number of result codes that can be returned as a result of errors during the request message routing/forwarding. All routing and forwarding errors are categorized as *protocol errors* and fall into the 3xxx class of the status codes. As a reminder, protocol errors use a specific "answer-message" Command Code Format (CCF) instead of the normal answer message for the request. The "E" error bit is also set in the "answer-message" command header. The protocol errors are handled on a hop-by-hop basis, which means that intermediate Diameter nodes may react to the received answer message. The intermediate node reacting to the error may try to resolve the issue that caused the error before forwarding the "answer-message" back to the downstream Diameter node.

`DIAMETER_REALM_NOT_SERVED`

(Status code 3003) Used when the realm of the request message is not recognized. This could be due to a lack of the desired realm in the routing table and the lack of a default route, or due to the requested realm being malformed.

`DIAMETER_UNABLE_TO_DELIVER`

(Status code 3002) Used in two situations:

- There is no Diameter node (either a server or an intermediate agent) in the destination realm that can support the desired application to which the request message belongs.
- The request message lacks the `Destination-Realm` AVP but has the `Destination-Host` AVP and the request should be routed. This case is covered in Section 4.3.

`DIAMETER_LOOP_DETECTED`

(Status code 3005) Used in situations where a Diameter node (typically an intermediate agent) notices that it has received a request message it had already forwarded. This implies that the Diameter network has a routing loop somewhere or that the DNS infrastructure has misconfigured zone files.

The Diameter node finds a loop by inspecting the routing AVPs in the received request message (discussed further in Section 4.3.2). In the case of routing loops, it makes little sense to attempt to re-route the request message, rather the Diameter node that detected the loop should raise an event to the network administration for further inspection and just return the answer message to the downstream node.

`DIAMETER_REDIRECT_INDICATION`

(Status code 3006) This status code is not a result of a routing error. It is used only in conjunction with redirect agents to redirect the request message and possibly subsequent request messages to a different Diameter node. The response is meant for the adjacent node, which should not forward it on. Recent updates (RFC 7075 [6]) to redirecting behavior added the ability to redirect a whole realm. A new status code was added for this purpose: `DIAMETER_REALM_REDIRECT_INDICATION` (status code 3011). However, this functionality works only for newly defined applications. Note that although not an error, the "E" bit is still set in the answer message.

`DIAMETER_APPLICATION_UNSUPPORTED`

(Status code 3007) Used in a situation where a Diameter request message reaches a Diameter agent that has no entry for the desired application in its routing table and thus the node cannot route/forward the request message.

4.5 Answer Message Routing

Answer messages always follow the Hop-by-Hop Identifier determined reverse path. When a Diameter mode receives an answer message, it matches the Hop-by-Hop Identifier in that message against the list of pending requests. If an answer does not match a known Hop-by-Hop Identifier, the node should ignore it. If the node finds a match, it removes the corresponding message from the list of pending requests and does the need processing for the message (e.g., local processing, proxying).

4.5.1 Relaying and Proxying Answer Messages

If the answer message arrives at an agent node, the node restores the original value of the Diameter header's Hop-by-Hop Identifier field and proxies or relays the answer message. If the last `Proxy-Info` AVP in the answer message is targeted to the local Diameter server, the node removes the `Proxy-Info` AVP before it forwards the answer message.

If the answer message contains a `Result-Code` AVP that indicates failure, the agent or proxy must not modify the `Result-Code` AVP, even if the node detected additional, local errors. If the `Result-Code` AVP indicates success, but the node wants to indicate an error, it can provide the appropriate error in the `Result-Code` AVP in the message destined towards the request message originator but must also include the `Error-Reporting-Host` AVP. The node must also send an STR on behalf of the request message originator towards the Diameter server.

4.6 Intra-Realm versus Inter-Realm Communication

RFC 6733 has an interesting statement in the request routing overview:

```
For routing of Diameter messages to work within an administrative
domain, all Diameter nodes within the realm MUST be peers.
```

The above implies a full mesh between all Diameter nodes within one realm. One can argue that a realm could be split into multiple administrative domains, however, since the realm is also piggybacked on the administration of the DNS, it is not obvious to claim that a "flat realm" could be more than one administrative domain.

For big operators, this full-mesh requirement is challenging to meet. Just think about the multi-million subscriber operator that has continent-wide geographical coverage where the network has to be partitioned because of operational and reliability reasons. Furthermore, the concept "all peers" would mean that only the peer table is consulted. However, the peer table has no application knowledge, therefore, even for pure peer connections, the routing table has to be consulted to determine the right peer connection for the desired application.

The `Destination-Host` AVP contains also the realm, since the value of the `Destination-Host` AVP is a FQDN. Therefore, it is possible to determine the destination realm even if the request message lacks the `Destination-Realm` AVP.

For standards compliance and to alleviate too big "flat realms", dividing a realm into multiple *sub-realms* is a valid solution. There, for example, the realm *example.com* would mean and point to "edge agents" of the realm. Anything inside the realm and also part of the host names (as seen in the `Origin-Host` AVP) would then contain more detailed realms such as *east.example.com* and *north.example.com*. The realm internal agents and routing/forwarding would then be based on these more detailed sub-realms to make them appear as multiple realms instead of a single flat realm. This approach is more or less analogous to DNS zone hierarchies.

4.7 Diameter Routing and Inter-Connection Networks

This section discusses the topic of large multi-realm inter-connecting Diameter networks. The discussion is not exhaustive, since the examples are limited only to few publicly known large deployments.

4.7.1 Inter-Connection Approaches

One of growing and expected to be huge Diameter inter-connection networks is the IPX network [10] serving cellular operators, and not just 3rd Generation Partnership Project (3GPP)-based cellular operators. 3GPP made a far-reaching decision to migrate all Signaling System 7 (SS7) based signaling interfaces with Diameter for the Evolved Packet System (EPS) in 3GPP Release 8. Eventually every Long-Term Evolution (LTE) enabled cellular operator has to use Diameter, not only to connect with their roaming partners, but also in their internal network.

GSM Association (GSMA) has been defining how inter-operator roaming works in practice. For EPS and LTE, GSMA produced the *LTE Roaming Guidelines* [11], which also detail the envisioned and recommended Diameter inter-operator network architecture. Figure 4.3 illustrates the "reference" inter-operator Diameter-based inter-connection network architecture.

The basic architectural approach is straightforward. Operators have relay agents, i.e., Diameter edge agents (DEA), on their network edges. These relay agents are then connected to internal agents, e.g., 3GPP specific Diameter routing agent (DRA) proxies or "vanilla" Diameter proxies with application-specific treatment of Diameter messages. Finally, the proxy agents are connected to the Diameter clients and servers. The connectivity within the operator's realm does not need to be a full mesh. However, for failover purposes, the peer connections within an operator realm are nearly a full mesh.

The connectivity between operators (and different realms) is realized using one or a maximum of two intermediate IPX roaming network providers. These IPX providers offer the transit of Diameter signaling traffic. The IPX provider may also deploy a number of intermediate agents, *IPX proxies*, for instance for value-added services. The IPX proxies can be relay agents only providing request routing services, or they can also be application-aware proxies doing application-level handling and/or manipulation of the transit traffic. Obviously, just deploying relay agents makes it easier to roll out new Diameter applications, since there is no need to upgrade intermediate agents in the IPX network for the new application support.

One of the obvious value-added services that IPX providers could offer is taking the burden of managing roaming partner connectivity relations and routing information on behalf on the customer operator. In this case, the customer operator only needs to do the following with its preferred IPX provider:

1. Inform the IPX provider with whom and which realms it has a roaming relationship.
2. Inform the IPX provider of any policies for filtering, Diameter level firewalling, correcting known anomalies in implementations, etc.
3. Route all traffic that goes outside its own realm(s) to the IPX provider's IPX proxy.
4. Receive all traffic coming through the IPX provider's IPX proxies.

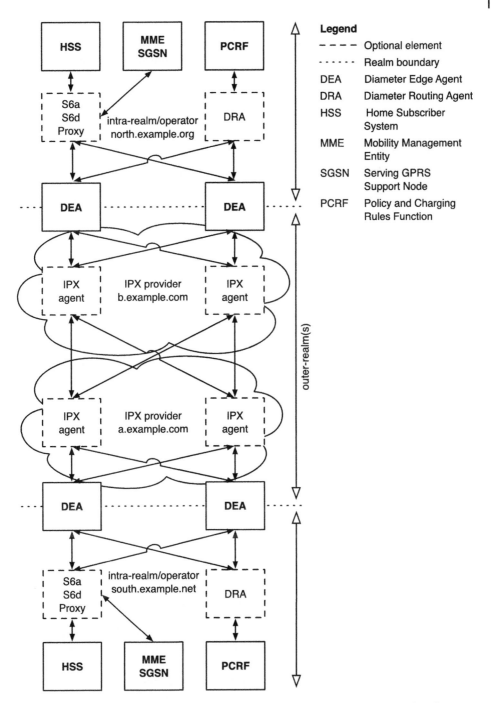

Figure 4.3 Example inter-connection Diameter network using the IPX-roaming network and IPX-proxies.

The above model simplifies greatly the operator's own Diameter routing bookkeeping and policy-based Diameter message processing at the network edges. The GSMA LTE Roaming Guidelines [11] also recommend using dynamic node discovery [2, 7] at the network edges or within IPX. The dynamic discovery eases the management of the next-hop peer discovery. Section 4.7.2 discusses the pros and cons of the dynamic node discovery in detail.

4.7.2 Dynamic Diameter Node Discovery

The Diameter routing infrastructure may form a complex topology due to the large number of roaming partners involved. The number of partners often exceeds hundreds of companies. The number of peers is therefore even larger, at least double the size, due to the recommendation of maintaining redundant peer connections for improved reliability. As a result, the routing tables for Diameter nodes in such a topology contain a large number of entries.

It is also common for each "foreign realm" to have a dedicated policy for handling and processing of request messages. In large Diameter networks even the internal realm topology can be complex. This results in a vast number of entries in the peer table, and, depending on the internal realm structure, also results in multiple sub-realm entries in the routing table. As an example, a realm `example.com` may have a sub-realm `sub.example.com`. All these contribute to the complexity and administrative overhead of the Diameter node operations and management. For scalability and managability reasons operators avoid storing comprehensive connectivity information to all internal as well as external nodes in each Diameter node. Default route entries and dynamic Diameter node discovery are useful tools to ease the deployment of large Diameter networks.

Similarly to the management of the DNS infrastructure in many enterprise networks, the internal and external views of the networks are kept separate. There is no need for realm-external nodes to learn the realm-internal topology or even Diameter node DiameterIdentities. It is also typically in the network administrators' interest to operate specific ingress into and egress points out of the network for security purposes. It's also not useful for a realm-internal node to dynamically discover realm-external nodes since direct peer connections outside realm are not allowed. Obviously the realm-internal DNS view could be configured so that all realm-internal DNS-based dynamic discovery attempts always resolve to the specific agents within the realm.

Therefore, in a typical large Diameter network deployment, the realm edge agents, which can be proxies or relays, are likely the ones initiating the discovery and also being populated into the public DNS to be discovered by realm-external nodes.

Example deployment architectures are illustrated below. In each architecture it is assumed that the Diameter client inside the originating realm has only a static route to the realm's edge agent. All traffic that exits the client's home realm is directed to the edge agents. Similarly, all traffic coming into the realm always goes through the edge agents. Direct connectivity between realm internal Diameter nodes is rarely if ever allowed in production networks. The reasons are the same as with the IP networking in general: better network control, manageability, and security.

4.7.2.1 Alternative 1

In Figure 4.4, the realm *foo.example.com* edge agent discovers the realm inter-connection network's edge agent when it tries to discover the "server B" Diameter node in the realm *bar.example.com*. Here, the realm *bar.example.com* DNS administration delegates the publishing of the edge agent DNS information to the inter-connection network provider.

4.7.2.2 Alternative 2

In Figure 4.5, the realm *foo.example.com* edge agent discovers the realm *bar.example.com* edge agent when it tries to discover the "server B" Diameter node in the realm *bar.example.com*. Here, the realm *bar.example.com* DNS administration publishes the edge agent DNS information in its own public DNS.

4.7.2.3 Alternative 3

In Figure 4.6, the realm *foo.example.com* edge agent only has a static "default" route to the inter-connection network's edge agent. The inter-connection network agent dynamically discovers the realm *bar.example.com* edge agent on behalf of the realm *foo.example.com* edge agent. Here, the realm *foo.example.com* has an agreement that the inter-connection network handles its realm-routing management, and that the *bar.example.com* DNS administration delegates the publishing of the edge agent DNS information in its own public DNS.

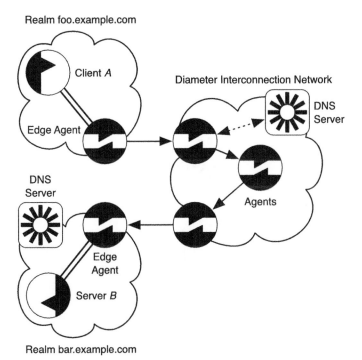

Figure 4.4 Dynamic peer discovery: delegating the publishing of DNS information.

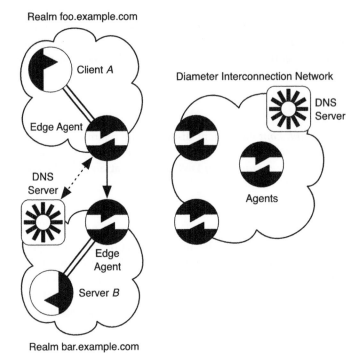

Figure 4.5 Dynamic peer discovery: realm publishes DNS information.

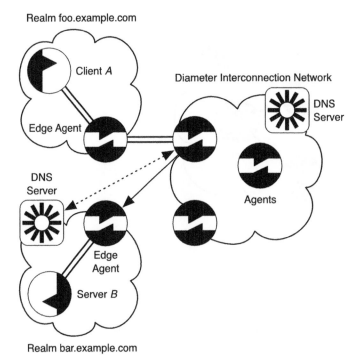

Figure 4.6 Dynamic peer discovery: static route between the client's edge agent and the inter-connection network.

4.8 Diameter Overload Control

Diameter Overload Control [12] is a recent larger solution concept done in the IETF and also adopted by 3GPP in their Diameter-based interfaces. For example, the 3GPP S6a interface [1] adopted Diameter overload control in Release 12. The basic architecture and the default functionality are described in the Diameter Overload Information Conveyance (DOIC) [13] specification. Figure 4.7 illustrates the high-level architecture of Diameter overload control. The main idea behind DOIC is to allow a message-receiving Diameter host or Diameter realm to inform message originator(s) that it is under a load condition. The message originators would then apply specific algorithms to back off and hopefully resolve the load condition.

There are a few key design ideas behind the Diameter overload control design:

1. Overload control information can be transported over any existing Diameter application that allows adding (piggybacking) arbitrary AVPs into its commands (i.e., has *[AVP] in its command's CCF).
2. The overload controlling algorithm is not fixed, and new algorithms and functionality can be added in a backward-compatible manner.
3. The solution is not tied to a specific architecture. The solution can work across realms, through agents (with DOIC awareness or not), and allow agents to represent other nodes for DOIC functions.

A good example of DOIC extensibility is the *load control* amendment [14] that was widely accepted in 3GPP Release 14 specifications. The load control adds a mechanism to convey *load information* (and make use of it) in addition to the mandatory to implement default *loss algorithm* specified in the DOIC specifications.

The DOIC specifies two roles for Diameter nodes: a *reporting node* (a message receiver) and a *reacting node* (a message originator). The reporting node sends periodic

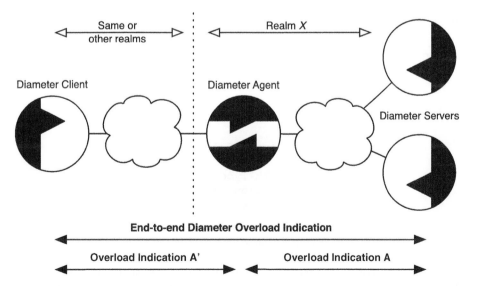

Figure 4.7 Simplified architecture choices for overload indication delivery for an arbitrary Diameter application.

updates of its overload condition (or a lack of it). The reacting node receives these reports and is supposed to react to the reported overload condition by applying the mutually agreed overload abatement algorithm. There are no predefined client-server roles in DOIC just like there are no such roles in Diameter. The "roles" are implicitly determined by the direction of the communication. The reporting and reacting nodes determine the identity (i.e., the DiameterIdentity) of their *"DOIC partner"* from the Origin-Host AVP or similarly the entire realm from the Origin-Realm AVP.[1]

A reacting node, which is the originator of the messages that may contribute to an overload condition on the receiving end, indicates its support for DOIC by including the OC-Supported-Features grouped AVP into every request message it originates (see Figure 4.8). The AVP includes the OC-Feature-Vector AVP, which in turn indicates one or more capabilities, e.g., the set of overload abatement algorithms the reacting node supports. The mandatory to support capability is the *loss algorithm* OLR_DEFAULT_ALGO. The reacting node also determines whether its communication counterpart supports DOIC from the possibly received OC-Supported-Features AVP.

If a reporting node determines the reacting node supports DOIC, it in turn indicates its support for DOIC by including the OC-Supported-Features AVP in response messages towards a reacting node. This OC-Feature-Vector AVP contains the set of mutually supported features. However, in a case of overload abatement algorithms, only a single mutually supported algorithm is returned out of possibly several candidates. The response messages towards the reacting node may also include the OC-OLR grouped AVP (see Figure 4.9). The OC-OLR AVP contains the actual *overload report (OLR)* information.

The DOIC specification does not detail the Diameter agent behavior or possible functions. Only the basic rules are laid out. Diameter Agents that support DOIC should relay all messages that contain the OC-Supported-Features AVP. An interesting function for a Diameter agent is to take the role of a reacting or reporting node for Diameter endpoints that do not support DOIC. Alternatively, a Diameter node may

```
OC-Supported-Features ::= < AVP Header: 621 >
                             [ OC-Feature-Vector ]
                         *  [ AVP ]
```

Figure 4.8 The Grouped OC-Supported-Features AVP and its sub-AVPs.

```
OC-OLR ::= < AVP Header: 623 >
              < OC-Sequence-Number >
              < OC-Report-Type >
              [ OC-Reduction-Percentage ]
              [ OC-Validity-Duration ]
          *   [ AVP ]
```

Figure 4.9 The Grouped OC-OLR AVP and its sub-AVPs.

---

1 The term "DOIC partner" is used in this chapter to refer to the DOIC endpoint at the other end of the DOIC communication, which may or may not be a direct Diameter peer. The term is not used in RFC 7683.

also add or reduce features to those advertised by DOIC-supporting nodes in their OC-Supported-Features AVP. In that case the Diameter agent also has to ensure consistency in its behavior with both upstream and downstream DOIC partners. Diameter agent overload and peer overload report amendment to the DOIC is a good example of a Diameter agent that actively participates in the Diameter overload condition handling [15].

4.8.1 Overload Reports

The OC-OLR AVP contains a set of fixed, mandatory sub-AVPs that are the same for all current and future abatement algorithms of the OLR. The optional sub-AVPs change depending on the supported and used abatement algorithms. Figure 4.9 illustrates the OC-OLR grouped AVP with sub-AVPs that are present with the default (must implement) loss abatement algorithm. The fixed AVPs are the OC-Sequence-Number and the OC-Report-Type AVPs. The OC-Sequence-Number AVP carries a monotonically increasing sequence number that DOIC partners (namely the reacting node) use to detect whether or not the contents of the OLR actually update the maintained overload control state (OCS) (covered in Section 4.8.2). The OC-Report-Type AVP informs the reacting node whether or not the contents of the OLR concern a specific node (the value of OC-Report-Type is HOST_REPORT) or an entire realm (the value of OC-Report-Type is REALM_REPORT).

The loss abatement algorithm is specified by the OC-Reduction-Percentage and the OC-Validity-Duration AVPs. The former indicates the percentage of the traffic that the reacting node is requested to reduce, compared to what it otherwise would send. The values between 0 and 100 are valid. The value 100 means that the reporting node will not process any received messages, and the value 0 means the overload condition is over (somewhat similar meaning as setting the OC-Validity-Duration to zero although the explicit ending of the overload condition should be signaled using the OC-Validity-Duration AVP). The OC-Validity-Duration AVP indicates how long the recently received OLR information is valid. The default value is 30 seconds with a maximum of 86,400 seconds. The value of zero (0) indicates the overload condition concerning this OCS state is over.

4.8.2 Overload Control State

Both reacting and reporting nodes maintain Overload Control State (OCS) for their active overload conditions. At the reacting node the active overload condition is determined from the received OC-OLR sub-AVP OC-Validity-Period with a non-zero value. At the reporting node the OCS state is created and maintained for DOIC partners when the overload condition is active.

The OCS states are indexed somewhat differently on the reacting and reporting nodes. A reacting node maintains an OCS entry for each Diameter *Application-Id + DiameterIdentity* tuple (Table 4.1). The DiameterIdentity is either a host from the Origin-Host AVP of the OLR (the OC-Report-Type AVP value is HOST_REPORT) or a realm from the Origin-Realm AVP of the OLR (the OC-Report-Type AVP value is REALM_REPORT).

Table 4.1 Overload control state for reacting nodes.

State information	Description
Sequence number	Detect whether the received `OC-OLR` updates the OCS state, i.e., if the received `OC-Sequence-Number` is greater than the stored sequence number, then update the OCS state. Otherwise, silently ignore the received `OC-OLR`.
Time of expiry	The validity time derived from the `OC-Validity-Duration` received in the OC-OLR. On an event of expiration the OCS entry for the overload condition is removed. Receiving the `OC-Validity-Duration` zero (0) signals immediate expiration of the overload condition.
Selected algorithm	The overload abatement algorithm selected by the reporting node and received from the `OC-Supported-Features` for the ongoing overload condition.
Per algorithm input data	Data specific to the implementation and the abatement algorithm. For example, in the case of the `OLR_DEFAULT_ALGO` this would include the `OC-Reduction-Percentage` from the OC-OLR.

A reporting node maintains an OCS entry for each Diameter *Application-Id +* DOIC partner *DiameterIdentity + supported abatement algorithm + report type* tuple (Table 4.2). The DiameterIdentity is always a host (from the `Origin-Host` AVP) of the request message that contained the `OC-Supported-Features`.

The complete information stored in an OCS state entry is an implementation decision. The original goal of the DOIC design was to maintain state only for overload conditions and respective "DOIC partners". However, it quickly turned out that this was not a realistic goal since the OCS ended up populated with sequence numbers and timers. The sequence numbers are used to check whether newly received `OC-OLR` AVPs and timers are used. Still, the maintained state is rather trivial.

Table 4.2 Overload control state for reporting nodes.

State information	Description
Sequence number	Last used sequence number with the latest `OC-OLR` sent to the DOIC partner. The sequence number is increased only when the intimation sent in the `OC-OLR` changes. For a new OCS state the sequence number is set to zero (0).
Validity duration	The validity time for the sent OLRs. When the overload condition ends the validity time is set to zero (0).
Expiration time	All sent OLRs have an expiration time in the reporting node's OCS state. The expiration time is equal to *current time* when the report was sent plus the validity duration.
Per algorithm input data	Implementation and abatement algorithm specific data. For example, in the case of the `OLR_DEFAULT_ALGO` this would include the `OC-Reduction-Percentage` from the OC-OLR.

4.8.3 Overload Abatement Considerations

Diameter did not have a mechanism to prioritize arbitrary messages over each other in a standardized and a generic manner until Diameter Routing Message Priority (DRMP) [16] was specified. Among several use cases enumerated in RFC 7944, the prioritization of messages when DOIC-originated overload abatement takes place is a prominent one. The DRMP instructs the Diameter nodes, for example, to make an educated throttling decision between different Diameter messages or a resource allocation.

The DRMP uses a single DRMP AVP in a Diameter message to indicate the relative priority of the message compared to other messages seen by a Diameter node. There are 16 priorities. It is important that priorities are assigned and managed in a coordinated manner within an administrative domain and even between domains/realms. Otherwise, the priorities can be mishandled or misinterpreted, or there could be too many messages with the same priority to realize any benefit of prioritization among messages.

References

1 3GPP. Evolved Packet System (EPS); Mobility Management Entity (MME) and Serving GPRS Support Node (SGSN) related interfaces based on Diameter protocol. TS 29.272, 3rd Generation Partnership Project, Jan. 2018.

2 J. Korhonen, M. Jones, L. Morand, and T. Tsou. Clarifications on the Routing of Diameter Requests Based on the Username and the Realm. RFC 5729, Internet Engineering Task Force, Dec. 2009.

3 C. Rigney, S. Willens, A. Rubens, and W. Simpson. Remote Authentication Dial In User Service (RADIUS). RFC 2865, Internet Engineering Task Force, June 2000.

4 B. Aboba, M. Beadles, J. Arkko, and P. Eronen. The Network Access Identifier. RFC 4282, Internet Engineering Task Force, Dec. 2005.

5 A. DeKok. The Network Access Identifier. RFC 7542, Internet Engineering Task Force, May 2015.

6 T. Tsou, R. Hao, and T. Taylor. Realm-Based Redirection In Diameter. RFC 7075, Internet Engineering Task Force, Nov. 2013.

7 V. Fajardo, J. Arkko, J. Loughney, and G. Zorn. Diameter Base Protocol. RFC 6733, Internet Engineering Task Force, Oct. 2012.

8 M. Jones, J. Korhonen, and L. Morand. Diameter Straightforward-Naming Authority Pointer (S-NAPTR) Usage. RFC 6408, Internet Engineering Task Force, Nov. 2011.

9 D. Mills, J. Martin, J. Burbank, and W. Kasch. Network Time Protocol Version 4: Protocol and Algorithms Specification. RFC 5905, Internet Engineering Task Force, June 2010.

10 GSMA. Guidelines for IPX Provider networks (Previously Inter-Service Provider IP Backbone Guidelines). Version 9.1, May 2013. http://www.gsma.com/newsroom/wp-content/uploads/2013/05/IR.34-v9.1.pdf.

11 GSMA. LTE and EPC Roaming Guidelines. Version 10.0, July 2013. http://www.gsma.com/newsroom/wp-content/uploads/2013/07/IR.88-v10.0.pdf.

12 E. McMurry and B. Campbell. Diameter Overload Control Requirements. RFC 7068, Internet Engineering Task Force, Nov. 2013.

13 J. Korhonen, S. Donovan, B. Campbell, and L. Morand. Diameter Overload Indication Conveyance. RFC 7683, Internet Engineering Task Force, Oct. 2015.

14 B. Campbell, S. Donovan, and J. Trottin. Diameter Load Information Conveyance. Internet-Draft draft-ietf-dime-load-09, Internet Engineering Task Force, 03 2022. Work in progress.

15 S. Donovan. Diameter Agent Overload and the Peer Overload Report. Internet-Draft draft-ietf-dime-agent-overload-11, Internet Engineering Task Force, 03 2022. Work in progress.

16 S. Donovan. Diameter Routing Message Priority. RFC 7944, Internet Engineering Task Force, Aug. 2016.

5

Diameter Security

5.1 Introduction

In the design of Internet protocols, security is approached in a structured way by analyzing threats, which requires a high-level understanding of the protocol's communication architecture [1] and has been described in earlier chapters, and then deriving security requirements. These security requirements can then be addressed by various security services, such as data integrity and confidentiality protection, authentication, authorization, etc. The IETF has developed building blocks offering such security services and offers guidance to specification authors on how to engineer security into protocols with RFC 3552 [2]. Guidance for considering privacy in protocol design is captured in RFC 6973 [3].

We approach a description of Diameter security in this book similarly to RFC 3552, in addition we provide background information about information security in Section 5.2. We discuss threats in Section 5.3, followed by a description of security services applied to the Diameter protocol in Section 5.4, and conclude with an example in Section 5.5.

Recall that historically a user would connect to a network by contacting a Network Access Server (NAS) over a land line via a modem, then had to authenticate, probably using a login and password, before being granted access to the Internet or a company network. The NAS then communicated with a back-end server and the AAA process took place. Since phone lines were not shared, the security of the network access was mainly provided by the AAA protocol, and in particular by the authentication interaction.

While dialup modems have disappeared, the basic principle of authenticating users before they are granted access to a network is still in widespread use. Mobile phones, for example, perform authentication via the subscriber identity module (SIM) and the authentication server checks whether the user is authorized before granting the user access to the mobile operator network. Another example is a remote access server (RAS) that enables company employees to access company internal resources from remote locations by establishing a virtual private network (VPN) connection, after the employee is successfully authenticated and authorized. In many deployments today, users access the network via wireless or cellular radio technology, and everyone within the transmission range can eavesdrop on the communication. Hence, there is not only the need to

Diameter: New Generation AAA Protocol – Design, Practice, and Applications, First Edition.
Hannes Tschofenig, Sébastien Decugis, Jean Mahoney and Jouni Korhonen.
© 2019 John Wiley & Sons Ltd. Published 2019 by John Wiley & Sons Ltd.

authenticate and authorize the access to the network, which is mainly of interest to the network operator, but also to secure the wireless link between the device and the NAS, such that unencrypted communication content is not available to an eavesdropper.[1]

The AAA architecture has therefore been designed to offer authentication and authorization of the device and user requesting access to the network and also to facilitate the distribution of keying material to enable integrity and confidentiality protection of data transmitted over the wireless link.

The authorization policies stored at AAA servers typically support a range of deployment scenarios. For example, in an enterprise WLAN environment, only employees may be entitled to access to the network, while at an Internet cafe access to the network and to the Internet may only be granted to paying customers.

The incentives for hackers to circumvent access control systems are obvious – in parallel, the techniques to protect the access control systems evolve rapidly to match the attacks as soon as new threats are identified. For example, the trend today in enterprise networks is not only to control access of remote connections in the RAS, but also to control all connections to the internal network, with the assumption that the protection granted by the enterprise premise's physical isolation is insufficient to protect against unauthorized access. For this purpose, architectures such as IEEE 802.1X [4] enforce authentication of devices connecting to wireless or wired [5] networks, similar to the remote access scenario. Here again, the AAA protocol is one component of the complete solution that secures the access to the network.

Besides network access control, Diameter is used nowadays in a variety of deployments where the nature of Diameter's signaling behavior is convenient for other applications and services as well. It is therefore important to understand the security properties Diameter offers when making a design decision to reuse Diameter for a specific service.

5.2 Background

For a better understanding of Diameter security concepts we will review a few basics. We do not aim, however, to provide a detailed treatment of cryptography, which is the study of mathematical techniques related to information security. There are many good books about cryptography and Internet security. For a good introduction to cryptography we recommend *Handbook of Applied Cryptography* by Alfred Menezes, Paul van Oorschot, and Scott Vanstone, which is available as a free download at http://cacr.uwaterloo.ca/hac/. For a better understanding of TLS we recommend *SSL and TLS: Designing and Building Secure Systems* by Eric Rescorla [6]. Eric Rescorla is co-author of the TLS and DTLS specifications and has co-chaired the IETF TLS working group for several years.

---

1 It is worth noting that communication security will most likely also be applied to communication interactions at higher protocol layers. For example, access to a web server using HTTP is likely to experience protection via TLS. This additional protection is necessary since the endpoints of the communication interaction are different and a website provider may be interested to use this communication security to protect their website login.

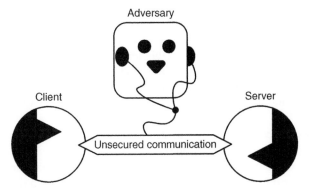

Figure 5.1 Adversary model.

For IPsec and IKEv2 we recommend the reader consults the original IETF specifications, namely [7] for IPsec ESP and [8] for IKEv2.

Many Internet communication protocols assume a very simple adversary model, which is shown in Figure 5.1. Two entities, a client and a server, interact with each other by exchanging messages over an unsecured communication channel. This communication channel is exposed to an adversary, which may eavesdrop, replay, reorder, delete, and inject messages.

Both the sender and the receiver want confidentiality of data communication so that an adversary cannot spy on the exchange and retrieve plaintext data. A receiver also wants authenticity of the exchanged data so that he has a guarantee that the data was sent by a certain sender and not injected or modified by an adversary. The obvious solution to this problem is to use a dedicated communication channel between the sender and the receiver that is physically isolated from any adversary. Needless to say, this is expensive and impractical for most purposes. Since it is more cost-effective to share communication channels, cryptography techniques are used to protect the data in transit over non-secure channels, such as Internet communications.

Well-established cryptographic algorithms, which are used to build cryptographic protocols, are known as cryptographic primitives. These cryptographic primitives can be split into three families of techniques with different properties, as described in the subsequent sub-sections.

5.2.1 Unkeyed Primitives

Unkeyed primitives, particularly hash functions, are used as one-way functions that map a large, variable-sized input message into a small fingerprint. A common size of such a fingerprint is 128 or 256 bits. Unkeyed primitives are typically used in combination with other techniques, such as digital signatures, to sign only a small message instead of an entire document. For hash functions the goal is to prevent an attacker from easily generating two messages that lead to the same fingerprint. Unkeyed primitives, as the name indicates, do not utilize any key as input.

An example of an unkeyed primitive is SHA-256 [9].

5.2.2 Symmetric Key Primitives

Symmetric key primitives are cryptographic functions that assume two parties[2] are in possession of a shared secret key, i.e., a bit string of variable length. To be useful this symmetric key has to be applied to a cryptographic function, such as a keyed message digest (also called message authentication code) or an encryption algorithm.

An example of a keyed message digest offering integrity services is HMAC-SHA256 [9].

When a keyed message digest is computed over a message M, often written as HMAC-SHA256(M, K), whereby K denotes the symmetric key, then it provides integrity protection and data origin authentication. Integrity protection ensures that the message receiver can verify whether message M has been modified in transit. When the key K is known only to the sender and the receiver, then the receiver can be sure that only a particular sender computed the message digest. An example of an encryption algorithm is the Advanced Encryption Standard (AES) [10]. Encryption algorithms, like AES, provide confidentiality services to ensure that messages transmitted by the sender can only be successfully decrypted by the intended recipient, who also knows the symmetric key. Since many encryption algorithms operate on blocks with a fixed length, the individual blocks need to be chained together so that their order cannot be rearranged. The mechanism of chaining the blocks together is also known as the mode of operation. An example of such a mode of operation is the Galois Counter Mode (GCM) [11].

5.2.3 Asymmetric Key Primitives

Unlike symmetric keys where the sender and the receiver share the same key, an asymmetric key primitive, also known as public key cryptography, requires that each party has two components: a public key and a private key. The private key is, as the name indicates, not shared with anyone whereas the public key is made available to everyone in an authentic fashion.

There are two main operations in public key cryptography: encryption and digital signature.

For the encryption operation, the sender uses the public key of the receiver to encrypt a message M. The encrypted message is then conveyed from the sender to the receiver, and only the corresponding private key, held by the receiver, can be used to obtain the original message M.

For the digital signature operation, the sender applies its own private key to the message M to create a digital signature of the message. The receiver verifies the integrity of the message M by using the sender's public key. If the verification is successful then the receiver can be certain that the message has been signed by the sender and that it has not been modified in transit.

An example of a popular public-key cryptosystem is RSA, named after the inventors Rivest, Shamir, and Adleman [12]. Since public key cryptography is computationally much more expensive than symmetric key cryptographic operations, many security

---

2 Symmetric key primitives are also used to secure multi-party communication protocols, and for some of these the symmetric key is shared between more than two parties. Since Diameter does not use multicast security, we will ignore this case.

protocols combine symmetric and asymmetric key primitives in a performance-efficient way. For example, public-key cryptography is used to authenticate the endpoints and to establish a fresh and unique session key. The established session key is then used with a symmetric key algorithm to secure the communication channel. Furthermore, in practical systems the above-described digital signature algorithm is typically combined with a hash function so that the digital signature is computed over the fingerprint instead of the original message. Likewise public key encryption is used in combination with a symmetric key cryptosystem so that the public key encryption is only applied to a randomly generated symmetric key and this symmetric key is used to encrypt the content instead.

While public key cryptography seems to offer huge benefits over its symmetric counterparts, it is worth pointing out that distributing public keys in an authentic manner among all communication parties can be challenging, particularly when the number of parties increases. For this purpose the public key infrastructure (PKI) has been developed, as described in more detail below.

Once an asymmetric algorithm has been used to distribute keying material for the protection of application data, it becomes difficult for an adversary to modify or inject data. One popular attack vector is therefore the generation or distribution of the keys: if the attacker can trick a communication partner into using its own public key, a technique called man-in-the-middle, then it can intercept the communication.

Hence, a logical follow-up question is how to ensure that a public key belongs to the intended party? This question relates to the concept of trust, which is an important topic in Diameter deployments as it reflects business relationships protected by business contracts (at least in most cases). We will provide a high-level overview in this chapter and a more detailed treatment in Section 5.5.

As described earlier, public-key cryptography enables the generation of digital signatures that can be generated only by the party in possession of the private key, and can be verified by anyone knowing the public key. As a result, if you have a public key that you know belongs to a legitimate party, any data you receive with a digital signature can be verified using that public key.

To bind an identifier and other attributes to the public key, the concept of a digital certificate was introduced. Table 5.1 lists several essential attributes contained inside a certificate.

Table 5.1 Essential attributes in a certificate.

Attribute	Description
Subject	Identifier for the entity for which the certificate was created. The issuer has verified the identifier of the subject.
Issuer	The certificate creator's identifier.
Validity period	Length of time the certificate is valid.
Issuer signature algorithm	Algorithm used by the issuer to compute a digital signature over the certificate.
Issuer's signature	Digital signature computed over the certificate.
Subject Public Key Information	Field containing the actual public key. The structure depends on the public key crypto-system used.

The issuer of a certificate creates a digital signature to protect the content of that certificate. This signature creates the binding between the public key and the remaining attributes (and the identifier of the subject in particular[3]). Without such a digital signature, everyone could modify the content of the certificate and would therefore be able to modify the association between the subject and the public key.

However, this digital signature created by the issuer leads to another problem: the digital signature created by the issuer needs to be verified again. Hence, the certificate of the issuer needs to be known as well. The issuer certificate may well be digitally signed by another issuer. To terminate the process, it is important to provision TLS and IKEv2 clients and servers with a trust anchor store, which contains the certificates of issuers (so-called root certificates or trust anchors). These trust anchors are typically self-signed certificates of certification authorities (CAs) but may well be certificates for any other party (in case of certificate pinning). It is therefore important for a software implementation to configure the trust anchor store with those certificates that are relevant for the given task.

To provide interoperability, the certificate format and other support protocols (e.g., those used for certificate revocation) require standardization, which has taken place in the ITU-T and the IETF. For example, the X.509 format [13] for encoding certificates has been used in TLS, the Internet Key Exchange protocol, and many other protocols. A more detailed description of certificates and the PKI can be found in *Planning for PKI* by Russ Housley and Tim Polk [14].

5.2.4 Key Length Recommendations

An important factor for determining the strength of symmetric and asymmetric cryptographic algorithms is the key size. The recommended length is based on several factors, including the best known attack against the algorithm family. For this reason, the recommended key length differs among the different algorithm families. Due to the increasing speed of computers and advances in cryptanalysis, key length recommendations (as well as the algorithms themselves) are adjusted from time to time.

Table 5.2 shows comparable key sizes for symmetric key algorithms and two categories of asymmetric algorithms. The values are taken from RFC 4492 [15]. The two categories of asymmetric algorithms are elliptic curve cryptography (ECC),[4] and the RSA and Diffie–Hellman algorithms. For example, the second row in the table reads as follows: a 112-bit symmetric key provides equivalent security to a 233-bit ECC key and a 2048-bit RSA key.

RFC 4492 provides a good illustration of what key sizes to expect when using different algorithms. Various publications, as summarized at http://www.keylength

---

3 In the PKI model the binding between the subject and the public key is established via the digital signature. There are other techniques for verifying such a binding, such as using out-of-band techniques where users manually assert the relationship. While these non-PKI-based techniques are fairly common in other communication protocols, such as instant messaging, network management, or Internet of Things, they are less commonly used in Diameter deployments.

4 For an introduction to ECC, we recommend *Guide to Elliptic Curve Cryptography* by D. Hankerson, A. Menezes, and S. Vanstone [16]. This book not only contains the theoretical foundations, but also offers algorithmic descriptions suitable for implementations that take performance optimizations into account. Cryptographic protocols based on ECC are also described, such as Elliptic Curve Digital Signature (ECDSA), which are used in protocols like TLS and IKEv2.

Table 5.2 Comparable key sizes (in bits).

Symmetric key	ECC	RSA/DH
80	163	1024
112	233	2048
128	283	3072
192	409	7680
256	571	15360

.com, offer key length recommendations. For example, key size recommendations for use with TLS-based ciphers can be found in [16] and the authors recommend Diffie–Hellman-based key lengths of at least 2048 bit, which corresponds to a 112-bit symmetric key and a 233-bit ECC key (according to the entries in Table 5.2). These recommendations are in line with those from other organizations, such as the National Institute of Standards and Technology (NIST) or the European Network and Information Security Agency (ENISA) [17]. The authors of [18] note that a symmetric 80-bit security level is sufficient for legacy applications for the coming years, but a symmetric 128-bit security level is the minimum requirement for new systems.

5.3 Security Threats

Information conveyed in Diameter is sensitive. It consists of personal data, such as geographic location (defined in RFC 5580 [19]), user identifiers (such as the Network Access Identifier (NAI) [20]), and account balance (specified in RFC 4006 [21]). It also includes security relevant information, such as keying material (e.g., RFC 7155 [22], RFC 4072 [23], RFC 6738 [24]), as well as authorization and service usage information (e.g., via accounting information). An adversary able to retrieve this data or inject new data will be able to gain information about end users (e.g., for surveillance purposes), interfere with services, reject access to services (as a denial-of-service attack), or commit fraud.

Since Diameter is versatile and usable in a variety of situations and applications with different security requirements, there are some intrinsic constants that help explain the security model of Diameter, starting with the actors involved in any Diameter deployment.

End User:
The end user of the service does not use the Diameter protocol directly, but utilizes a service provider. The authentication interactions may be transparent for the user, for example when it is performed by hardware security modules without the user's interaction, like a mobile phone's subscriber identity module, or they may require user involvement (e.g., by having the user enter a PIN or password). For our discussion, unless specified otherwise, we will not distinguish between the human end user and his or her device, such as a mobile phone or tablet.

Service Provider:
The service provider offers a specific service for users, such as Internet access, voice over IP (VoIP) services, or IPTV, and incurs costs in terms of bandwidth, service operation, hardware, etc. The service provider operates the client-side of the AAA protocol as part of the service. Note that the service provider may offer network access as a service (which is the case of the classical use of network access authentication utilized by an Internet Service Provider (ISP)), or the service provider may offer application layer services, such as a VoIP service. It is also possible for an ISP to offer application layer services.

Identity Provider:
The identity provider operates the server side of the AAA protocol and of the Diameter application. The identity provider is responsible for providing the user with credentials, processing requests forwarded from the service provider, authenticating end users, making access control decisions and forwarding authorization information to the service provider, collecting accounting records, etc. For commercial services, the accounting records are used to determine the actual service usage and consequently the service charge.

AAA Broker:
The AAA broker is an intermediary that connects service providers and identity providers in larger eco-systems. Not all deployments involve AAA brokers, but many large-scale deployments utilize these intermediaries in order to scale their business relationships, i.e., to lower the number of contracts needed between identity providers and service providers. The functions of an AAA broker are to establish and maintain relationships between different parties (technically as well as business-wise), to route Diameter messages, to offer security protection and privacy, to deal with fraud, to ensure that resource usage is correctly metered, and to perform financial settlement. While many AAA brokers do not publish a description of their architecture and processes, educational federations have been more transparent. An example of such an educational federation (using RADIUS) is Eduroam and a description of their infrastructure can be found in [25].

Each of these actors has different views on what is considered valuable data and how it should be protected. End users value the protection of their identity and privacy, and want to be charged correctly for their service usage. The service provider needs to keep accurate records of the service usage for billing and planning purposes. Keeping data about the service usage may also be a legal obligation for providing the service. The identity provider needs to protect the subscriber database against unauthorized access and data loss. AAA brokers are interested in ensuring the correct flow of accounting data to prevent fraud. It is worth noting that for a single end user service, several Diameter servers may be involved, with different Diameter applications, as is the case in an Long-Term Evolution (LTE) network.

There are different adversaries to the Diameter infrastructure that need to be considered. An adversary aims to gain unauthorized access to data or attempts to inject or forge data within the AAA system. We have to assume that the entities bound by contract for the delivery of the service are trusted to fulfill their functions, meaning they will not purposely exploit the data collected for purposes other than AAA functions. In the security context, the term "trust" often has a negative connotation since trusting parties requires more entities to behave appropriately, which typically increases the attack surface. Of course, not all data in Diameter is required to be known by every entity since some Diameter payloads have only end-to-end significance and do not need to be processed by routing agents. However, certain Diameter payloads must be visible to all Diameter entities along the path, as explained in earlier chapters.

From the point of view of the end user, an adversary can:

- Passively observe and collect data at any point along the entire communication chain (from the device to the network operator, during the service initialization and usage), in order to extract useful information such as identifiers, credentials, meta-data about the user's communication partners, communication and mobility patterns, etc. or to replay this exchange to gain access to the service at a later time. In the context of government surveillance activities this is indeed a real threat.
- Impersonate the service provider in order to trick the end user into sending any information that can be reused by this adversary, or at least allow this adversary to gain access to the service. This might lead to higher resource consumption, which may increase the service charge.
- Impersonate end users to prevent them from accessing the service, or to cause the early termination of the service. This may, for example, allow the adversary to get faster Internet access or more bandwidth in a resource-shared environment.
- Impersonate end users in order to get access to the service, possibly after the legitimate user has ended using the service.

From the point of view of a service provider, an adversary can:

- Impersonate an end user or otherwise try to gain access to the service, for example by exploiting flaws in the access control system.
- Eavesdrop on AAA message exchanges for surveillance purposes.
- Inject itself as a Diameter node in order to gain access to Diameter messages. This gives an adversary full control over the Diameter protocol, including access to the Diameter message content, the ability to modify messages, to inject new messages, and to control Diameter clients.
- Disrupt the normal operation of the service (e.g., via a denial-of-service attack) in order to blackmail a service provider or to damage their reputation.
- Replay previous authentication exchanges with a backend server in order to re-authenticate a device without involving the end user.
- Compromise the equipment of the network operator in order to get remote access to this device and to change the behaviour of the software running on this equipment.

From the point of view of an identity provider, an adversary can:

- Disrupt the flow of the accounting records in order to use the service without being charged.
- Compromise the AAA server to gain access to the subscriber data and other sensitive data stored in databases or in memory.
- Mount a denial-of-service attack to make the server inaccessible.

From the point of view of a AAA broker, an adversary can:

- Disrupt the routing of Diameter messages and thereby impact the service of the broker.
- Gain unauthorized access to data stored by the AAA broker, such as accounting records.

- Compromise the security of the AAA broker to launch an attack against service and identity providers.
- Inject Diameter messages to cause fraudulent transactions.

All the above-described threats can be categorized into three categories:

(1) passive attacks, where an adversary simply collects and analyzes information without interfering;
(2) active attacks on a communication channel, where an attacker may change the content of the packets exchanged between two parties involved in the system; and
(3) active attacks on Diameter nodes participating in the system.

A short note about the threat of traffic analysis, which is a technique where an adversary is able to learn useful information of even encrypted communication interactions, for example by observing patterns in the exchange, the frequency of exchanges, unencrypted header information, or even the lack of communication. Diameter is less vulnerable to traffic analysis since Diameter links are typically encrypted and busy with messages pertaining to a large number of users and possibly many Diameter applications. Individual sessions belonging to a single user are therefore well hidden among the other Diameter messages when Diameter links are properly protected with channel security.

5.4 Security Services

In this section, we first present the security measures built into the Diameter protocol and how they address some of the threats described in the previous section. Subsequently, we discuss possible ways to mitigate generic security threats that are also applicable to Diameter deployments.

5.4.1 Diameter Security Model

As described in Chapter 2, the base Diameter specification reuses existing security protocols and exchanged data is cryptographically protected at two levels:

1. On a Diameter hop-by-hop basis, Diameter messages are carried between peers over network protocols that provide security features, such as mutual authentication of the peers, data confidentiality and integrity, and replay protection.
2. On an end-to-end basis, a Diameter AVP may contain cryptographically protected data. Such end-to-end protection is, however, not commonly used by Diameter applications. It is worth noting that the first version of Diameter, specified in RFC 3588 [26], provided a dedicated flag in the AVP header, the "P" flag, to indicate when the AVP needed to be encrypted for end-to-end security. This flag was deprecated in the revision of the standard, since the initial attempt to standardize Diameter end-to-end security was never completed. Only recently has a new attempt to standardize an end-to-end security solution for Diameter been restarted, see [27].

While the Diameter specification mandates that hop-by-hop security is provided by a lower layer protocol, it offers the option to use either IPsec, combined with IKE or IKEv2 for key management, or preferably TLS or DTLS. We will discuss these alternatives below. Note that it is important to compare the identities exchanged by these security protocols with the DiameterIdentity – the same way a web browser compares the URL entered by a user with the identity received in the certificate sent by the server during the TLS exchange. A browser issues a security alert in case of mismatch – and to make sure that the remote peer, even when presenting valid credentials, is authorized to join your Diameter network. RFC 6125 [28] goes into great detail on how to verify and represent application service identities within the Internet PKI using X.509 (PKIX) certificates in the context of TLS.

5.4.1.1 Secure Transports

Modern communication security protocols operate in two phases: a computationally expensive handshake phase followed by a phase that makes use of symmetric key cryptography to efficiently protect application data. The latter phase consists of the data exchanged by the endpoints of the protocol – the payload from the point of view of the protocol – where the security properties, such as confidentiality or authenticity, are cryptographically enforced. The handshake phase is dedicated to the setup and management of the cryptographic keying material and parameters for the bulk data protection, such as mutual authentication of the endpoints, key derivation, negotiation and establishment of the security parameters, rekeying, etc. At the transport layer the Transport Layer Security (TLS) protocol is widely used nowadays to protect HTTP exchanges on the web, commonly referred to as HTTPS – technically, HTTP over TLS. TLS is a layered protocol that consists of several sub-protocols, including a handshake protocol and a record layer protocol. TLS can only be used over a transport protocol that provides reliable and ordered delivery of the data, such as TCP. To also offer communication security services for unreliable transport protocols like UDP, the Datagram Transport Layer Security (DTLS) protocol was designed, which is strongly inspired by TLS. SCTP is another transport protocol, younger than TCP and UDP. SCTP is a reliable protocol – all data sent on one side will be delivered on the other side, or an error reported – that provides to applications fine control over the ordering of the data delivered: data chunks can be sent unordered, as in UDP, or ordered within streams that are independent, as if parallel TCP connections were established, but with less overhead. In addition, SCTP supports multiple IP addresses for each endpoint, which makes it more resistant to network failures. The initial version of Diameter in RFC 3588 [26] mandated TLS to be used to protect SCTP links. In that case a separate TLS context must be established for each SCTP stream, which is highly resource consuming. RFC 6733 [29] resolved this issue by promoting DTLS over SCTP instead of TLS. For the use of TCP, the choice of TLS remained unchanged. SCTP is the transport protocol of choice for Diameter as its ordering granularity and resilience to network failures are interesting properties for AAA traffic. However, it is also a complex and relatively new protocol (at least compared to TCP), and the configuration of firewalls may be more complex. The combination of TLS over TCP is more commonly deployed than DTLS over SCTP in the industry today. From this point on, we will use the term (D)TLS when referring to either TLS over TCP or DTLS over SCTP, for the sake of simplicity.

RFC 3588 stated that, if a Diameter connection was not protected by IPsec, then the CER/CEA exchange needs to include an `Inband-Security-Id` AVP with a value of TLS. The TLS handshake would begin after completion of the CER/CEA exchange.

This behavior has been deprecated by RFC 6733. However, TLS may still be negotiated via this `Inband-Security-Id` AVP to support older Diameter implementations. RFC 6733 introduced another important change with the allocation of a dedicated port for (D)TLS-protected connections. In RFC 3588, a single port was used for all connections, and the TLS protection was started only after the capability exchange step, through negotiation. This method opened the door to downgrading attacks, where an attacker would be able to trick the Diameter peers to not use TLS. As a result, the revised method uses a dedicated port where (D)TLS handshake takes place immediately after connection establishment and before the Diameter capability exchange. The Diameter capability exchange is therefore protected as well. With this design the Diameter identity of the remote peer may not be known at the time of the (D)TLS handshake. Therefore, a Diameter implementation needs to store the authenticated identity and to compare this value with the identity presented during capability exchange. A connection needs to be terminated in case of mismatch. Failure to enforce this binding would enable a Diameter peer to impersonate another peer, possibly from another domain. Diameter can be used over a non-secure transport protocol, TCP or SCTP, if the security is implemented at the IP layer, using IPsec ESP [7]. IPsec ESP offers bulk data encryption and a separate handshake protocol, namely Internet Key Exchange version 2 (IKEv2). From the point of view of communication security, IPsec+IKEv2 and (D)TLS offer the same security services but the design details of the two protocols vary considerably. It is, however, worth noting that (D)TLS security is implemented in applications in modern systems, while IPsec comes as a system service, for example for management of system-wide VPN connections. Therefore, choosing IPsec in a Diameter deployment might, from a software design point of view, create additional challenges for enforcing the binding between the authenticated identity in the IKEv2 exchange and the Diameter Identity presented by a peer. Furthermore, when using SCTP with its multi-homing support, the configuration of IPsec may become challenging.

5.4.1.2 Authorization

It is not sufficient to ensure that a remote peer presents a valid credential, and an explicit authorization check must be performed on all Diameter peers to make sure that each peer, with credential matching its Diameter identity, is allowed to join the Diameter network.

For a single administrative domain this authorization check can be implemented, for example, by whitelisting peers in configuration files. It becomes a maintenance burden when such whitelists have to be extended to connections that span separately administered domains. Using a dedicated public key infrastructure for the Diameter network is more efficient: a specific certification authority (CA) issues certificates to be used only by Diameter peers in the domain, independent of certificates used by other services. This works reliably when all Diameter peers are configured to accept certificates signed only by this dedicated CA, the trust anchor for the Diameter domain, and accept no other CA certificates – most operating systems are configured with a long list of trust anchors that need to be ignored by a Diameter stack.

As with any PKI-based model, it is important for an implementation to ensure that

- Path validation for certificates in IKEv2 or (D)TLS stacks is performed,
- The validity for all certificates in the chain between the CA certificate and the certificate presented by the remote party is checked,

- X.509v3 extensions are verified, in particular that all intermediate CA certificates are authorized to issue certificates,
- The revocation status of all certificates in the chain, using a certificate revocation list (CRL) or an Online Certificate Status Protocol (OCSP) responder [30], is verified.

Fortunately, these security checks are usually performed by the security protocol stack, and a Diameter implementation generally needs only to enforce the binding between the identity used for authentication and the Diameter Identity.

5.4.2 Relation to Threats

The security model of Diameter described previously aims to provide the following assurances:

- The communication between two Diameter peers is cryptographically protected. This protection includes authentication, confidentiality, replay, and integrity protection.
- All peers joining the Diameter network are legitimate entities involved in the business relationship covering the provision of a service. This insurance is provided partly by the mutual authentication at the cryptography level, and partly by the explicit authorization step.

The threats identified in Section 5.3 related to end users are not addressed by these security services. This is somewhat expected since the end user (or the device of the end user) is not directly involved in the Diameter exchange. The threats related to passive attackers are properly addressed, provided that cryptography is configured with appropriate algorithms and key lengths. The threats related to active attacks against the communication channel are addressed as well, provided that the mutual authentication and authorization of the Diameter peers is performed.

5.4.3 Mitigating Other Threats

By following the security recommendations from this chapter, the possible attacks to Diameter links, passive or active, are mitigated. Remaining from our initial threat description are the attacks on the Diameter endpoints themselves and on the end user of the service.

Attacks on Diameter endpoints encompass both remote and physical attacks on the equipment where the Diameter stack is running. The target can be the Diameter stack itself, the operating system where this stack is running, or any other entry point on this system. The mitigation of such a threat is not specific to Diameter. The following rules help to limit the risks:

1. Reduce the attack surface by limiting the number of services running on the entities that run Diameter.
2. Install security patches and install OS and Diameter stack security updates from reliable sources.
3. Limit the access from the outside world using a firewall; only Diameter ports should be open.
4. For Diameter stack developers, follow practices of secure software development. This includes penetration testing, peer review of source code, and static code analysis.

5. For nodes that store user-specific data ensure protection against unauthorized access.
6. Limit the physical access to the equipment as much as possible.
7. Protect the private key corresponding to the certificate used for mutual authentication. Make sure the key is generated in a secure way – if an attacker can predict the key, cryptography is useless [31].
8. Rotate the key periodically and follow the latest recommendations in terms of key lengths and algorithms.

The mitigation of the threats to the end user is dependent on the service and the network access technology. In the case of IEEE 802.11i [32], for example, EAP is used during the authentication step, allowing mutual authentication between the end user device and the service provider. At the end of a successful authentication, a key is generated and transported from the service provider to the network operator equipment over AAA protocol. This key is used to bootstrap the security of the channel between the end user's device and the network equipment, effectively binding this security to the authentication and preventing all the threats that we described on a shared access medium such as WiFi.

In addition to these considerations, there are more intrinsic methods that help prevent information leakage, mostly of interest for Diameter application developers. For example, RFC 4962 [33] describes how key material must be handled in an AAA protocol to limit the security risks, and RFC 6973 [3] gives good recommendations for preserving the privacy in Internet protocols.

5.5 PKI Example Configuration in `freeDiameter`

In this section we explain how to secure `freeDiameter` using the guidance provided in this chapter. The `freeDiameter` implementation, similar to many commercial products, supports TLS protection of Diameter communication by using GnuTLS, an open source TLS implementation. Many TLS stacks are designed with the use of securing Web traffic in mind and hence they typically offer a wide range of features.[5]

The ability to protect SCTP using TLS, as specified in RFC 3588, is offered by `freeDiameter`. However, this approach requires a lot of resources when a large number of streams are used. The alternative, namely to protect SCTP with DTLS, is work in progress in `freeDiameter` and not available at the time of writing.

5.5.1 The Configuration File

The good news is that configuring TLS in `freeDiameter` is easy since there is only a handful of configuration parameters. For extra customization we describe the `freeDiameter` configuration file in detail in Section B.7.

---

5 Note that the current version of `freeDiameter` only supports a subset of the GnuTLS capabilities. For example, it is not possible to use a private key stored in a hardware security module or a trusted execution environment. Hardware support for storing keying material and for executing cryptographic operations provides an additional layer of defense when the Diameter stack or the operating system is compromised.

Certificate and Private Key:

Each Diameter peer needs to possess a certificate along with the private key since we use asymmetric cryptography in our example. The most important configuration parameter is therefore the path to this X.509 certificate and to the corresponding private key. This key pair will be used to authenticate the Diameter peer. The path is specified with the `TLS_Cred` parameter. It is important to note that the CN (Common Name) field of the certificate must match the DiameterIdentity of the peer. Both files should be PEM-encoded. Protecting the private key with a password is currently not supported in `freeDiameter`.

Trust Anchors:

The path to one or more trust anchor(s) is configured using the `TLS_CA` parameter. The trust anchor should be PEM-encoded and by default the trust anchor store is empty. Only incoming TLS connections authenticated with a certificate anchored in one of the specified trust anchors will be accepted. Of course, other security checks may lead to a certificate verification failure, for example if the DiameterIdentity of a peer does not match the common name in the certificate, or if the certificate has expired.

Certificate Revocation Lists:

`freeDiameter` has support for certificate revocation lists (CRLs) that help protect the network when a certificate is revoked, for example when the private key is compromised. The `TLS_CRL` parameter specifies the path to the CRL.[6] This configuration can be omitted if the CRL feature is not used.

Ciphersuites:

The `TLS_Prio` parameter enables fine control over the TLS ciphersuite selection. The TLS ciphersuites concept defines a combination of cryptographic algorithms, key sizes, and key exchange methods.

To display the configuration file, type the following command on server.example.net:

```
$ cat /home/freediameter/freeDiameter/test/freeDiameter.conf
```

The configuration file has the following content:[7]

```
(...)
###################################################################
TLS_Cred = "/home/freediameter/freeDiameter/test/cert.pem",
           "/home/freediameter/freeDiameter/test/privkey.pem";
TLS_CA = "/home/freediameter/freeDiameter/test/ca.pem";
ConnectPeer = "client.example.net" { No_SCTP; };
LoadExtension = "dbg_msg_dumps.fdx" : "0x1020";
# LoadExtension = "test_app.fdx";
```

In this configuration we are not using the CRL feature. The `ConnectPeer` parameter serves three purposes:

- authorizes a peer with identity client.example.net to join the Diameter network,

---

6 Note that there are several possible improvements for the `freeDiameter` implementation. For example, modifing the CRL file requires the `freeDiameter` process to be restarted. Supporting OCSP is another useful feature enhancement, particularly when used with the TLS OCSP stapling functionality. OCSP allows the certificate status to be determined online. OCSP stapling allows OCSP responses to be included in the TLS handshake, which reduces latency and the separate communication interaction with the OCSP server.

7 Note that the exact configuration will depend on which example scripts were run previously.

- establishes a connection to this peer periodically, and
- uses TCP for connection attempts.

5.5.2 The Certificate

The certificate shown below can be found in the `test` directory on server.example.net (look for cert.pem), and is an example of what is described in Table 5.1 of Section 5.2. Here we show an X.509v3 certificate with the serial number 3 that was issued by the certification authority with the name freediameter.example.net based on information found in the Common Name (CN) of the `Issuer` attribute. Unless specified explicitly, the DiameterIdentity is automatically configured by `freeDiameter` from the FQDN. This certificate is valid from 3 February 2014 at 14:56:49 GMT to 1 February 2024 at 14:56:49 GMT. The `Subject` contains the identity of this server (server.example.net) in the CN attribute. The `Subject Public Key Info` field specifies the RSA parameters. Various X.509v3 extensions are followed by the digital signature computed by the issuer. The `Basic Constraints` field indicates that this is not a CA certificate, i.e., it cannot be used by server.example.net to issue further certificates. The `Key Usage` field indicates permissions, namely the use of the key for digital signatures, key encipherment, and non-repudiation. Code signing, for example, is not allowed with this certificate.

———————————————————— Certificate ————————————————————

```
Certificate:
    Data:
        Version: 3 (0x2)
        Serial Number: 3 (0x3)
    Signature Algorithm: sha1WithRSAEncryption
        Issuer: CN=freediameter.example.net
        Validity
            Not Before: Feb  3 14:56:49 2014 GMT
            Not After : Feb  1 14:56:49 2024 GMT
        Subject: CN=server.example.net
        Subject Public Key Info:
            Public Key Algorithm: rsaEncryption
                Public-Key: (1024 bit)
                Modulus:
                    00:ac:48:db:a6:a7:2d:4f:86:fd:db:58:22:0d:b6:
                    33:d5:e6:80:f5:1f:c8:75:a6:98:c5:34:08:02:f1:
                    c9:23:d6:27:e5:f6:cf:87:7d:3c:b9:4a:cf:6f:88:
                    f1:2c:ed:bb:05:db:d9:ef:f9:e9:4f:41:5d:0e:4c:
                    97:14:93:e4:15:4e:95:1c:69:50:00:7d:21:70:dd:
                    ca:4a:6e:aa:a4:98:77:eb:0f:db:c8:aa:9c:92:50:
                    e3:c7:a1:fe:2c:3a:c7:f3:45:62:30:6b:47:25:e3:
                    6d:38:9b:87:7d:f8:e6:f5:ef:08:8c:fc:ee:46:19:
                    87:e6:58:42:d9:dd:8d:13:09
                Exponent: 65537 (0x10001)
        X509v3 extensions:
```

```
X509v3 Basic Constraints:
    CA:FALSE
X509v3 Key Usage:
    Digital Signature, Non Repudiation,
    Key Encipherment
X509v3 Subject Key Identifier:
    C7:AE:A0:EF:04:72:4A:8E:EB:47:A3:7A:9C:0E:51:
    D3:2D:67:DF:BB
X509v3 Authority Key Identifier:
    keyid:5F:86:3C:1D:7E:EC:13:9F:B1:E3:B8:B6:50:
    6D:FB:47:EE:A3:9E:F5

Signature Algorithm: sha1WithRSAEncryption
    aa:c6:0a:e9:81:52:1e:e4:5b:99:2b:ee:ac:2c:fc:d3:e8:0a:
    f1:9f:6c:10:36:19:f0:d2:e3:bc:45:f5:82:93:18:f8:da:1a:
    2c:71:1c:33:42:98:40:4a:cc:49:b2:df:36:e3:49:e0:cb:88:
    f7:6d:d5:89:3f:5d:97:14:4b:6d:a1:fe:cf:9e:fc:4a:20:b8:
    e6:a8:1a:27:55:a0:b3:00:26:0f:53:76:6c:3f:f6:82:e9:f5:
    4a:1b:23:69:d8:f8:a5:86:f6:69:8e:42:e6:7b:92:57:85:dc:
    d0:2f:7d:34:5d:f4:96:7e:4b:35:d6:3e:64:f3:d4:0e:aa:a1:
    f8:d3 [ Another 128 bytes skipped ]
```

5.5.3   Protecting Exchanges via TLS

This section offers hands-on experiments to show TLS with different configuration settings.

To run these tests you have to set up the virtual environment, as described in Section A.6, and client.example.net and server.example.net must be ready to use. The instructions in Section A.6 prepare the appropriate environment.

1. Prepare the network configuration profile for the server.example.net VM:

```
$ nw_configure.sh server.example.net
```

2. Configure `freeDiameter` for the TLS experiment:

```
$ fD_configure.sh 5_srv_s
```

3. Display the FQDN of server.example.net:

```
$ hostname --fqdn
server.example.net
```

4. Display the certificate subject name to ensure it matches the FQDN:

```
$ openssl x509 -in ~/freeDiameter/test/cert.pem -subject -noout
subject= /CN=server.example.net
```

5. Run `freeDiameter`:

```
$ freeDiameterd
(...)
12:22:56  NOTI   freeDiameterd daemon initialized.
```

6. Configure and start the client with the following commands on client.example.net:

```
$ nw_configure.sh client.example.net
$ fD_configure.sh 5_cli_s
$ freeDiameterd
(...)
12:20:12  NOTI   freeDiameterd daemon initialized.
```

When we start the daemon on the client machine, a connection is successfully established and the two peers start exchanging periodic messages:

```
(...)
12:22:56  NOTI   freeDiameterd daemon initialized.
12:22:56  NOTI   SND to 'client.example.net': 'Capabilities-Exchange-
    Request'0/257 f:R--- src:'(nil)' len:152
12:22:56  NOTI   RCV from 'client.example.net': (no model)0/257 f:----
    src:'client.example.net' len:164
12:22:56  NOTI   'STATE_WAITCEA'            -> 'STATE_OPEN' 'client
    .example.net'
12:23:27  NOTI   RCV from 'client.example.net': (no model)0/280 f:R---
    src:'client.example.net' len:80
12:23:27  NOTI   SND to 'client.example.net': 'Device-Watchdog-
    Answer'0/280 f:---- src:'(nil)' len:92
```

From the debugging output in the log file it is difficult to determine whether the connection is actually TLS protected. An easy way to check is to run Wireshark, a network protocol analyzer that captures protocol traffic from a network interface.

7. Open another terminal window on the server.example.net VM.
8. In the terminal window, type the following command to verify whether the connection is encrypted:

```
$ sudo tshark -d tcp.port==5868,ssl -f "port 5868 or port 3868"
```

You may see an error mentioning the programming language Lua, and a warning about running Wireshark as the root user. Since we are not using any Lua scripting with Wireshark, we can ignore the warning. When a DWR/DWA exchange occurs, you will see the following on the screen:

```
Capturing on eth0
 31.939257 192.168.35.5 -> 192.168.35.10 TLSv1.2 311 Application Data
 31.940489 192.168.35.10 -> 192.168.35.5 TLSv1.2 439 Application Data
 (...)
```

Wireshark is only able to show the unencrypted parts of the TLS 1.2 handshake.

9. To stop the exercise, type Ctrl-C in the terminal windows running freeDiameter and Wireshark.

5.5.3.1 Common Name and Hostname Mismatch

If the content of the Subject: CN attribute given in the certificate does not match the hostname used in freeDiameter, then the certificate will be rejected by free-Diameter. To verify this failure scenario, enter the following commands on the server.example.net machine, which will change the CN to client.example.net:

1. Reconfigure server.example.net for this experiment:

```
$ fD_configure.sh 1_cli
```

2. Verify that the FQDN of server.example.net is indeed server.example.net:

```
$ hostname --fqdn
server.example.net
```

3. Note that its CN in the certificate is client.example.net:

```
$ openssl x509 -in ~/freeDiameter/test/cert.pem -subject -noout
subject= /CN=client.example.net
```

4. Start `freeDiameter`:

```
$ freeDiameterd
```

The following information will be displayed:

```
12:14:44  NOTI   libfdproto '1.2.0-1256(bd6b40c9f731)' initialized.
12:14:44  NOTI   libgnutls '2.12.14' initialized.
12:14:44  NOTI   libfdcore '1.2.0-1256(bd6b40c9f731)' initialized.
12:14:44  ERROR  TLS: Local certificate '/home/freediameter/
    freeDiameter/test/cert.pem' is invalid :
12:14:44  ERROR    - The certificate hostname does not match
    'server.example.net'
12:14:44  ERROR  ERROR: in '((fd_conf_parse()))' :        Invalid
    argument
12:14:44  ERROR  ERROR: in '(fd_core_parseconf(conffile))' :    Invalid
    argument
12:14:44  FATAL! Initiating freeDiameter shutdown sequence (1)
12:14:44  NOTI   freeDiameterd framework is stopping...
12:14:44  NOTI   Shutting down server sockets...
12:14:44  NOTI   Sending terminate signal to all peer connections
12:14:44  NOTI   server.example.net: Going to ZOMBIE state (no more
    activity)
12:14:44  NOTI   'STATE_CLOSED' -> STATE_ZOMBIE (terminated)     'server
    .example.net'
12:14:44  NOTI   Waiting for connections shutdown... (16 sec max)
```

5.5.3.2 Unprotected Exchanges

While it is strongly advised to use TLS protection for operating a Diameter network, it is sometimes convenient to disable the use of TLS during the software development to see protocol exchanges over the wire. This may simplify debugging in certain situations.

`freeDiameter` was designed to encourage users to configure TLS; as a result it is slightly more difficult to disable TLS than to enable it. The TLS configuration has to be performed globally, as we have seen above. It can be disabled for each connection individually by editing the configuration file.

It is not necessary to hand-edit the configuration file for this exercise. We have created a configuration for you that can be applied to the VMs.

1. Apply the configuration settings to the server machine:

```
$ fD_configure.sh 5_srv_ns
```

2. Apply the configuration settings to the client machine:

```
$ fD_configure.sh 5_cli_ns
```

3. To see that TLS has been disabled, view the configuration file on the client machine:

```
$ cat /home/freediameter/freeDiameter/test/freeDiameter.conf
(...)
################################################################
TLS_Cred = "/home/freediameter/freeDiameter/test/cert.pem",
           "/home/freediameter/freeDiameter/test/privkey.pem";
TLS_CA = "/home/freediameter/freeDiameter/test/ca.pem";

ConnectPeer = "server.example.net" { No_TLS; No_SCTP; };

LoadExtension = "dbg_msg_dumps.fdx" : "0x1020";
# LoadExtension = "test_app.fdx";
```

4. Run freeDiameter on both the server and client machines:

```
$ freeDiameterd
(...)
13:11:19  NOTI    freeDiameterd daemon initialized.
13:11:35  NOTI    RCV from '<unknown peer>': (no model)0/257 f:R---src:
    '(nil)' len:164
13:11:35  NOTI    SND to 'client.example.net': 'Capabilities-Exchange-
    Answer'0/257 f:---- src:'(nil)' len:164
13:11:35  NOTI    No TLS protection negotiated with peer
    'client.example.net'.
13:11:35  NOTI    'STATE_CLOSED' -> 'STATE_OPEN' 'client.example.net'
13:12:05  NOTI    SND to 'client.example.net': 'Device-Watchdog-Request
    '0/280 f:R--- src:'(nil)' len:80
13:12:05  NOTI    RCV from 'client.example.net': (no model)0/280 f:----
    src:'client.example.net' len:92
```

The log now contains a message highlighting that TLS is not used with that peer.

5. Open another terminal window on the server machine and run Wireshark again:

```
$ sudo tshark -d tcp.port==5868,ssl -f "port 5868 or port 3868"
```

We can now see the Diameter messages clearly:

```
(...)
 75.616937 192.168.35.10 -> 192.168.35.5 DIAMETER 146 cmd=Device-
    WatchdogRequest(280) flags=R--- appl=Diameter Common Messages(0)
    h2h=32f91958 e2e=e258c34f
 75.617501 192.168.35.5 -> 192.168.35.10 DIAMETER 158 cmd=Device-
    WatchdogAnswer(280) flags=----- appl=Diameter Common Messages(0)
    h2h=32f91958 e2e=e258c34f
```

If you do not see these messages and Wireshark instead prints only TCP requests and responses then Wireshark needs to be told to interpret these messages as Diameter payloads. Please consult the Wireshark manual for the "decode as" functionality.

6. To stop the exercise, type Ctrl-C in the terminal windows running freeDiameter and Wireshark.

5.5.3.3 Certificate Revocation

The next example shows how the CRL list specified in the configuration file can be used to communicate certificates that have been revoked.

1. Configure the client.example.net machine as follows:

```
$ nw_configure.sh client.example.net
$ fD_configure.sh 5_cli_r
```

2. Verify that the CRL is configured:

```
$ cat /home/freediameter/freeDiameter/test/freeDiameter.conf
(...)
##################################################################
TLS_Cred = "/home/freediameter/freeDiameter/test/cert.pem",
           "/home/freediameter/freeDiameter/test/privkey.pem";
TLS_CA = "/home/freediameter/freeDiameter/test/ca.pem";
TLS_CRL = "/home/freediameter/freeDiameter/test/crl.pem";

No_SCTP;

ConnectPeer = "server.example.net";

LoadExtension = "dbg_msg_dumps.fdx" : "0x1020";
# LoadExtension = "test_app.fdx";
```

Let us assume that the private key stored on server.example.net has been compromised. Consequently, the CA has to revoke the corresponding certificate. Subsequently, a new certificate would be created and the revoked certificate is added to the CRL list.

3. Configure the server.example.net machine to use a certificate that has been revoked. An attacker gaining access to the compromised private key would in effect present himself with a similar setup to join the Diameter network:

```
$ nw_configure.sh server.example.net
$ fD_configure.sh 5_srv_r
```

4. Check the certificate being used:

```
$ cat /home/freediameter/freeDiameter/test/freeDiameter.conf
(...)
##################################################################
TLS_Cred = "/home/freediameter/freeDiameter/test/revoked.pem",
           "/home/freediameter/freeDiameter/test/privkey.pem";
TLS_CA = "/home/freediameter/freeDiameter/test/ca.pem";

ConnectPeer = "client.example.net";

LoadExtension = "dbg_msg_dumps.fdx" : "0x1020";
# LoadExtension = "test_app.fdx";
No_SCTP;
```

5. Run `freeDiameter` on both the client and server machines.

```
$ freeDiameterd
```

We can confirm that the connection is rejected on the client.example.net machine:

```
(...)
13:49:05  NOTI    freeDiameterd daemon initialized.
13:49:05  ERROR   TLS: Remote certificate invalid on socket 13 (Remote:
    'server.example.net')(Connection: '{---T} TCP
    ,#13->192.168.35.10(5658)') :
13:49:05  ERROR    - The certificate is not trusted (unknown CA?
    expired ?)
13:49:05  ERROR    - The certificate has been revoked.
13:49:05  ERROR   ERROR: in '(fd_tls_verify_credentials(conn->
    cc_tls_para.session, conn, 1))' :    Invalid argument
```

```
13:49:05  ERROR  ERROR: in '(fd_cnx_handshake(cnx, GNUTLS_CLIENT, 1,
    peer->p_hdr.info.config.pic_priority, ((void *)0)))' :
    Invalid argument
```

6. To stop the exercise, type Ctrl-C in the terminal windows running `freeDiameter` and Wireshark.

As an exercise, you can confirm that removing the `TLS_CRL` parameter from the client configuration file and then restarting the test will allow the two peers to successfully establish a connection.

For a real-world deployment it is important to think about how the case of compromised keys would be handled, including how impacted customers and users would be notified. CRLs and OCSP are two standardized mechanisms offering a method of certificate revocation. They are, however, not a solution for modifying the trust anchor store. Modifying the trust anchor store can be accomplished using software updates, network management protocols, or even using the Trust Anchor Management Protocol (TAMP) [34]. In any case, a product has to provide an answer for revoking certificates and for modifying the trust anchor store.

5.6   Security Evolution

Internet security is constantly evolving: new security threats arise, security solutions change, and cryptanalysis advances. The recommendations we give in this book are based on the state of the art in 2018, therefore the reader is strongly encouraged to consult the IETF DIME working group pages [35] for the latest specifications, information on our website at https://diameter-book.info for updates, and new releases of the `freeDiameter` implementation.

For recommendations on the use of new ciphersuites in TLS consult the IETF working group. In addition, the Internet Engineering Research Task Force (IRTF) Crypto Forum Research Group (CFRG) at https://irtf.org/cfrg is a valuable resource for guidance on algorithms.

References

1 E. Rescorla and IAB. Writing Protocol Models. RFC 4101, Internet Engineering Task Force, June 2005.
2 E. Rescorla and B. Korver. Guidelines for Writing RFC Text on Security Considerations. RFC 3552, Internet Engineering Task Force, July 2003.
3 A. Cooper, H. Tschofenig, B. Aboba, J. Peterson, J. Morris, M. Hansen, and R. Smith. Privacy Considerations for Internet Protocols. RFC 6973, Internet Engineering Task Force, July 2013.
4 802.1X-2010 – IEEE Standard for Local and Metropolitan Area Networks: Port-Based Network Access Control, May 2010.
5 802.1AE-2006 – IEEE Standard for Local and Metropolitan Area Networks: Media Access Control (MAC) Security, June 2006.
6 E. Rescorla. *SSL and TLS: Designing and Building Secure Systems*. Addison Wesley, 2001.

7 S. Frankel and S. Krishnan. IP Security (IPsec) and Internet Key Exchange (IKE) Document Roadmap. RFC 6071, Internet Engineering Task Force, Feb. 2011.

8 C. Kaufman, P. Hoffman, Y. Nir, P. Eronen, and T. Kivinen. Internet Key Exchange Protocol Version 2 (IKEv2). RFC 7296, Internet Engineering Task Force, Oct. 2014.

9 D. Eastlake 3rd and T. Hansen. US Secure Hash Algorithms (SHA and SHA-based HMAC and HKDF). RFC 6234, Internet Engineering Task Force, May 2011.

10 National Institute of Standards and Technology. FIPS 197, Advanced Encryption Standard (AES), pages 1–51, Nov. 2001.

11 M. Dworkin. Recommendation for Block Cipher Modes of Operation: Galois/-Counter Mode (GCM) and GMAC, National Institute of Standards and Technology SP 800- 38D, Nov. 2007.

12 R. Rivest, A. Shamir, and L. Adleman. A Method for Obtaining Digital Signatures and Public-Key Cryptosystems. *Communications of the ACM*, 21:120–126, 1978.

13 D. Cooper, S. Santesson, S. Farrell, S. Boeyen, R. Housley, and W. Polk. Internet X.509 Public Key Infrastructure Certificate and Certificate Revocation List (CRL) Profile. RFC 5280, Internet Engineering Task Force, May 2008.

14 R. Housley and T. Polk. *Planning for PKI: Best Practices for Deploying Public Key Infrastructure*. Wiley, 2001.

15 S. Blake-Wilson, N. Bolyard, V. Gupta, C. Hawk, and B. Moeller. Elliptic Curve Cryptography (ECC) Cipher Suites for Transport Layer Security (TLS). RFC 4492, Internet Engineering Task Force, May 2006.

16 D. Handerson, A. J. Menezes, and S. Vanstone. *Guide to Elliptic Curve Cryptography*. Springer, 2004.

17 ENISA. Algorithms, Key Sizes and Parameters Report – 2013, Oct. 2013. http://www.enisa.europa.eu/activities/identity-and-trust/library/deliverables/algorithms-key-sizes-and-parameters-report.

18 Y. Sheffer, R. Holz, and P. Saint-Andre. Recommendations for Secure Use of Transport Layer Security (TLS) and Datagram Transport Layer Security (DTLS). RFC 7525, Internet Engineering Task Force, May 2015.

19 H. Tschofenig, F. Adrangi, M. Jones, A. Lior, and B. Aboba. Carrying Location Objects in RADIUS and Diameter. RFC 5580, Internet Engineering Task Force, Aug. 2009.

20 B. Aboba, M. Beadles, J. Arkko, and P. Eronen. The Network Access Identifier. RFC 4282, Internet Engineering Task Force, Dec. 2005.

21 H. Hakala, L. Mattila, J.-P. Koskinen, M. Stura, and J. Loughney. Diameter Credit-Control Application. RFC 4006, Internet Engineering Task Force, Aug. 2005.

22 G. Zorn. Diameter Network Access Server Application. RFC 7155, Internet Engineering Task Force, Apr. 2014.

23 P. Eronen, T. Hiller, and G. Zorn. Diameter Extensible Authentication Protocol (EAP) Application. RFC 4072, Internet Engineering Task Force, Aug. 2005.

24 V. Cakulev, A. Lior, and S. Mizikovsky. Diameter IKEv2 SK: Using Shared Keys to Support Interaction between IKEv2 Servers and Diameter Servers. RFC 6738, Internet Engineering Task Force, Oct. 2012.

25 K. Wierenga, S. Winter, and T. Wolniewicz. The eduroam Architecture for Network Roaming. RFC 7593, Internet Engineering Task Force, Sept. 2015.

26 P. Calhoun, J. Loughney, E. Guttman, G. Zorn, and J. Arkko. Diameter Base Protocol. RFC 3588, Internet Engineering Task Force, Sept. 2003.

27 H. Tschofenig, J. Korhonen, G. Zorn, and K. Pillay. Security at the Attribute-Value Pair (AVP) Level for Non-neighboring Diameter Nodes: Scenarios and Requirements. RFC 7966, Internet Engineering Task Force, Sept. 2016.

28 P. Saint-Andre and J. Hodges. Representation and Verification of Domain-Based Application Service Identity within Internet Public Key Infrastructure Using X.509 (PKIX) Certificates in the Context of Transport Layer Security (TLS). RFC 6125, Internet Engineering Task Force, Mar. 2011.

29 V. Fajardo, J. Arkko, J. Loughney, and G. Zorn. Diameter Base Protocol. RFC 6733, Internet Engineering Task Force, Oct. 2012.

30 S. Santesson, M. Myers, R. Ankney, A. Malpani, S. Galperin, and C. Adams. X.509 Internet Public Key Infrastructure Online Certificate Status Protocol – OCSP. RFC 6960, Internet Engineering Task Force, June 2013.

31 D. Eastlake 3rd, J. Schiller, and S. Crocker. Randomness Requirements for Security. RFC 4086, Internet Engineering Task Force, June 2005.

32 IEEE. 802.11i-2004 – IEEE Standard for Information Technology – Telecommunications and Information Exchange Between Systems – Local and Metropolitan Area Networks – Specific Requirements Part 11: Wireless LAN Medium Access Control (MAC) and Physical Layer (PHY) Specifications Amendment 6: Medium Access Control (MAC) Security Enhancements, July 2004.

33 R. Housley and B. Aboba. Guidance for Authentication, Authorization, and Accounting (AAA) Key Management. RFC 4962, Internet Engineering Task Force, July 2007.

34 R. Housley, S. Ashmore, and C. Wallace. Trust Anchor Management Protocol (TAMP). RFC 5934, Internet Engineering Task Force, Aug. 2010.

35 IETF. Diameter Maintenance and Extensions (DIME), Mar. 2014. http://datatracker .ietf.org/wg/dime/charter/.

6

Diameter Applications

6.1 Introduction

This chapter discusses some widely used Diameter applications. Recall that application messages are exchanged between Diameter nodes only after capabilities have been discovered.

We will start with the application defined by the base specification, Base Accounting (Section 6.2), then look at another accounting application defined by the IETF called Credit-Control (Section 6.3). Next, we will take a look at an authorization application to specify quality of service (Section 6.4). The section that covers the authentication application, Diameter Extensible Authentication Protocol (EAP), also has a `freeDiameter` example (Section 6.5). The last and most complex application to be covered in this section is an authentication and authorization application used in mobile networks called the S6a Interface (Section 6.6).

Note that this chapter provides overviews of the applications' modes of operation, commands, and AVPs. The finer details for each application can be found in the application's RFC or technical specification.

We will cover some examples of these applications as they are extended by a standards development organization known as the 3rd Generation Partnership Project (3GPP) and deployed in mobile networks. The work of 3GPP is a good example of the adoption of Diameter and the development of Diameter applications. The area of mobile telecommunications provides fertile ground for Diameter applications due to the size of the networks, the heavy interoperability process performed by vendors and operators, and feedback from operational fields.

Over 30 Diameter applications have been defined and used in 3GPP networks. 3GPP refers to these applications as interfaces or reference points between two network elements and provides them with short labels, S6a, Rf, Gx, etc. 3GPP groups their specifications under a release system, which, when frozen, provides a stable platform upon which to implement the specifications. As of this writing Release 14 is the most current, frozen release.

6.2 Base Accounting

Accounting is the collection of information on resource usage, occurring in parallel with a user activity such as accessing the network or placing a voice call, for the purpose of

Diameter: New Generation AAA Protocol – Design, Practice, and Applications, First Edition.
Hannes Tschofenig, Sébastien Decugis, Jean Mahoney and Jouni Korhonen.
© 2019 John Wiley & Sons Ltd. Published 2019 by John Wiley & Sons Ltd.

Table 6.1 Base accounting application.

Reference	RFC 6733 [1]
Application-Id value	3
Commands	Accounting-Request/Answer (ACR/ACA) 271

billing, capacity planning, auditing, or cost allocation. Accounting takes place after a user has been authenticated and authorized for a service.

The Diameter base specification defines an accounting application, known as the base accounting application (Table 6.1), that is not a standalone application but is used by other Diameter applications, such as authorization applications, that also want to provide accounting information.

In Diameter, accounting information is not streamed, but is instead sent in discrete messages. Depending on the type of user access and the direction received from the accounting server, an accounting client can send accounting information via just one message, which contains an *event* record, or via multiple messages, which start with a *start* record, continue with *interim* records, and end with a *stop* record. The accounting server then needs a way to correlate the records it receives with the application and the user.

6.2.1 Actors

The actors for this application include the Diameter application that is using the base Accounting-Request and Accounting-Answer commands in addition to its own Diameter application commands for authentication and authorization. We will call this application a Network Access Server (NAS). The other actors include the accounting application client and the accounting application server.

6.2.2 Accounting Application Setup

Typically a successful transaction with the NAS starts the accounting application. The accounting application server selects how often the accounting application client should send accounting records based on provisioned information about the user and any relationships with roaming partners (if the accounting application is deployed in a mobile network), and communicates this mode to the client via the `Acct-Interim-Interval` and `Accounting-Realtime-Required` AVPs in its main application response.

`Acct-Interim-Interval`: The inclusion of this AVP (type Unsigned32) instructs the client to send snapshots of usage during a user's session, known as an interim record. Interim records provide a partial accounting of a user's session in case a device or network problem prevents the delivery of a session summary message or session record.
- The omission of this AVP or its inclusion with its value field set to 0 means that the client should produce only event records, start records, and stop records.

- The inclusion of the AVP with its value field set to a non-zero value means that the client produces interim records between the start record and the stop record. The value field of this AVP specifies the interval in seconds between the sending of these records. The client starts a timer when it sends a start record, sends the first interim record roughly when this interval time elapses, and continues to send interim records at timed intervals until the session ends, at which point the client sends a stop record. The client randomizes the production times of interim records to avoid creating message storms around a common service start time.

Accounting-Realtime-Required: This AVP provides guidance to the client on what the client should do when its transfer of accounting records is delayed or unsuccessful. Real-time accounting is the processing of information on resource usage within a defined time window in order to limit financial risk.

1. DELIVER_AND_GRANT: The service is granted only when there is a connection to an accounting server or to a set of accounting servers.
2. GRANT_AND_STORE: Service is granted when there is a connection or for as long as the client can store records. This is the default behavior if the Accounting-Realtime-Required AVP is not included.
3. GRANT_AND_LOSE: Service will be granted even if the records cannot be delivered or stored.

Note that the Accounting-Realtime-Required AVP doesn't indicate whether the accounting data is actually sent to the server in real time. The Diameter Credit-Control Application (Section 6.3) and the 3GPP Rf interface (Section 6.2.10) are examples of real-time accounting applications.

When the client receives a successful authentication and/or authorization message from the Diameter server, it starts collecting accounting information for the session according to the server's directives.

Depending on which *accounting service*, described in the next section, is used by the Diameter application, the client sends accounting information either to the main application server or to a centralized accounting server. The client transmits accounting records to the server via the Accounting-Request message. The server acknowledges with the Accounting-Answer message.

If the client interacts with an accounting server directly after successful authorization, then that accounting server can override directives given previously by the main application server by including the Acct-Interim-Interval or Accounting-Realtime-Required AVP in the Accounting-Answer message. The client always uses the latest values received.

6.2.3 Accounting Services

Diameter applications that want to incorporate the base accounting commands can route them in one of the following ways:

Split Accounting Service
The Accounting-Request and Accounting-Answer commands carry the Application-Id of 3, which is the Application-Id of the Diameter base accounting application, and

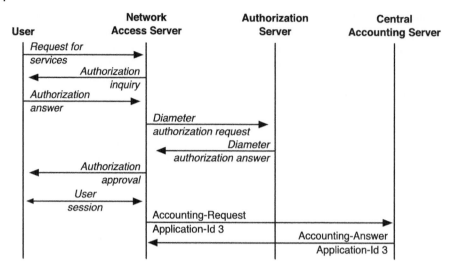

Figure 6.1 Split accounting service. The user's protocol is abstracted, as is the Diameter authorization application.

they can be routed to Diameter nodes other than the server that handles the main Diameter application, such as centralized accounting servers (Figure 6.1). Since the Application-Id doesn't provide information about the specific Diameter application that is sending the commands, the accounting server needs to inspect the message's other AVPs or their content to correlate messages.[1]

Because of this overhead, if you are creating a new Diameter application that requires an accounting component (designing new applications is covered in Chapter 7), the split accounting service is not recommended.

Examples of applications that use split accounting services are the Diameter Extensible Authentication Protocol (EAP) Application (Section 6.5) and the 3GPP Rf interface (Section 6.2.10).

Coupled Accounting Service

The Accounting-Request and Accounting-Answer commands carry the Application-Id of the Diameter application that is using it. The accounting messages are routed the same way as the other Diameter application messages and thus go to the authorization server (Figure 6.2). The authorization server then either processes the accounting records itself or forwards them to an accounting server.

When creating new Diameter applications that have an accounting component, the best current practice is to use the coupled accounting service.

Examples of applications that use a coupled accounting service are the Diameter Network Access Server Application [2] and the Diameter Mobile IPv4 Application [3].

---

1 You may have noticed the `Acct-Application-Id` AVP, which is specified by the base protocol, is not mentioned as a way to differentiate applications when using the split accounting service. The `Acct-Application-Id` AVP is used when the `Vendor-Specific-Application-Id` is present, and it must be included if the vendor-specific application has an accounting component. However, the base accounting application does not have any vendor-specific AVPs.

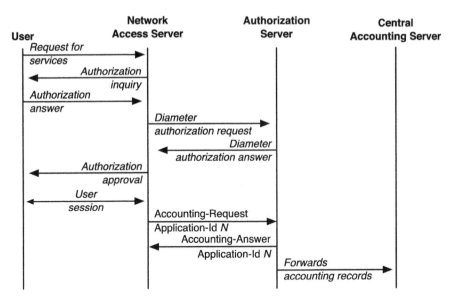

Figure 6.2 Coupled accounting service. The user's protocol is abstracted, as is the Diameter authorization application.

6.2.4 Accounting Records

Accounting records, which summarize the user's resource consumption, vary by the type of accounted service and the authorization server's directions. The client sends the accounting record in an Accounting-Request, where the details are captured in the service-specific AVPs.

If the accounted service is a one-time event, meaning that the start and stop of the event are simultaneous, then the client sets the `Accounting-Record-Type` AVP to `EVENT_RECORD` in the Accounting-Request.

If the accounted service session continues for some length of time, then the client sets the `Accounting-Record-Type` AVP to the value `START_RECORD` in the initial Accounting-Request for a session. When the client sends the last Accounting-Request, it sets the `Accounting-Record-Type` AVP value to `STOP_RECORD`. If the authorization server has enabled interim accounting, then the client sends additional records between the `START_RECORD` and `STOP_RECORD`, marked `INTERIM_RECORD`.

A session sequence then is either one record with `Accounting-Record-Type` AVP set to the value `EVENT_RECORD` or several records starting with `START_RECORD`, followed by zero or more `INTERIM_RECORD`s and a single `STOP_RECORD`. Since a session can have more than one accounting sequence, the `Accounting-Sub-Session-Id` is used to identify a particular sequence.

6.2.5 Correlation of Accounting Records

A Diameter application defines the concept of a session that is being accounted, and it may define the concept of a multi-session. For instance, the Diameter Network Access Server Application [2] treats a single point-to-point (PPP) connection [4] to an NAS as one session and a set of multilink PPP sessions as one multi-session.

An application can correlate accounting records within a session by using the Session-Id AVP found in the messages. If the accounting messages use a different Session-Id than the application sessions, then other session-related information is needed to perform correlation.

When an application requires multiple accounting sub-sessions, then each sub-session is differentiated by an Accounting-Sub-Session-Id AVP. The Session-Id remains constant for all sub-sessions and is used to correlate all the sub-sessions to a particular application session.

There are cases where an application needs to correlate multiple application sessions into a single accounting record; the accounting record may span multiple Diameter applications and sessions used by the same user at a given time. When the main application server determines that a request belongs to an existing session, it sends the Acct-Multi-Session-Id AVP during the authorization phase. The client then includes the Acct-Multi-Session-Id AVP in all subsequent accounting messages. The contents of Acct-Multi-Session-Id AVP are implementation specific, but are globally unique and do not change during the life of a session.

When the split accounting service is used, a centralized Diameter accounting server will need to sort messages from different applications; however, the Accounting-Request messages all use the same Application-Id of 3. The accounting server will need to look for application-specific AVPs in order to determine to which application the message applies.

6.2.6 Sending Accounting Information

The client sends an Accounting-Request (Figure 6.3) command to provide accounting information to the main application server or to an accounting server. In addition to the AVPs covered below, Accounting-Request messages usually include service-specific accounting AVPs.

By issuing an Accounting-Request corresponding to the authorization response, the client or the local realm implicitly agrees to provide the service indicated in the authorization response. If the local realm cannot provide the service, it sends a DIAMETER_UNABLE_TO_COMPLY error message within the Accounting-Request.

The server acknowledges an Accounting-Request command with an Accounting-Answer command (Figure 6.4). The Accounting-Answer command contains the same Session-Id as the corresponding request.

In addition to the AVPs listed in RFC 6733, Accounting-Answer messages usually include service-specific accounting AVPs.

6.2.7 Accounting AVPs

This section describes the AVPs that carry accounting usage information related to a specific session.

Accounting-Record-Type: Identifies the type of accounting record:
1 EVENT_RECORD: This record contains all information relevant to the service, and it is the only record of the service.
2 START_RECORD: Initiates an accounting session and contains accounting information that is applicable to the start of the session.

```
<ACR> ::= < Diameter Header: 271, REQ, PXY >
             < Session-Id >
             { Origin-Host }
             { Origin-Realm }
             { Destination-Realm }
             { Accounting-Record-Type }
             { Accounting-Record-Number }
             [ Acct-Application-Id ]
             [ Vendor-Specific-Application-Id ]
             [ User-Name ]
             [ Destination-Host ]
             [ Accounting-Sub-Session-Id ]
             [ Acct-Session-Id ]
             [ Acct-Multi-Session-Id ]
             [ Acct-Interim-Interval ]
             [ Accounting-Realtime-Required ]
             [ Origin-State-Id ]
             [ Event-Timestamp ]
          *  [ Proxy-Info ]
          *  [ Route-Record ]
          *  [ AVP ]
```

Figure 6.3 Accounting-Request Command Code Format.

```
<ACA> ::= < Diameter Header: 271, PXY >
             < Session-Id >
             { Result-Code }
             { Origin-Host }
             { Origin-Realm }
             { Accounting-Record-Type }
             { Accounting-Record-Number }
             [ Acct-Application-Id ]
             [ Vendor-Specific-Application-Id ]
             [ User-Name ]
             [ Accounting-Sub-Session-Id ]
             [ Acct-Session-Id ]
             [ Acct-Multi-Session-Id ]
             [ Error-Message ]
             [ Error-Reporting-Host ]
             [ Failed-AVP ]
             [ Acct-Interim-Interval ]
             [ Accounting-Realtime-Required ]
             [ Origin-State-Id ]
             [ Event-Timestamp ]
          *  [ Proxy-Info ]
          *  [ AVP ]
```

Figure 6.4 Accounting-Answer Command Code Format.

3 INTERIM_RECORD: Contains cumulative accounting information for an existing session. Interim Accounting Records are sent every time a re-authentication or re-authorization occurs or when triggered by application-specific events. The use of interim records is determined by the Acct-Interim-Interval AVP.

4 STOP_RECORD: Ends an accounting session and contains cumulative accounting information relevant to the existing session. When a service uses the accounting portion of the Diameter protocol, even in combination with an application, session termination messages are not used, since a session is terminated with an accounting stop message.

Accounting-Record-Number: Identifies this record within one session. Since Session-Id AVPs are globally unique, the combination of Session-Id and Accounting-Record-Number AVPs can be used to identify a single record within a session.

Acct-Session-Id: This AVP is only used when RADIUS/Diameter translation occurs. This AVP contains the contents of the RADIUS Acct-Session-Id attribute.

Acct-Multi-Session-Id: Links multiple, related accounting sessions, where each session has a unique Session-Id but the same Acct-Multi-Session-Id AVP. The server provides the value for this AVP, which the client includes in all accounting messages for the given session.

Accounting-Sub-Session-Id: Contains the accounting sub-session identifier. The combination of the Session-Id and this AVP is unique per sub-session, and the value of this AVP is monotonically increased by one for all new sub-sessions. If the client does not include this AVP, then there are no sub-sessions in use, except when Accounting-Record-Type is set to STOP_RECORD, which signals the termination of all sub-sessions for a given Session-Id.

6.2.8 freeDiameter Example

Here is an example of a Split Accounting Service message exchange captured from freeDiameter:

```
'Accounting-Request'
  Version: 0x01
  Length: 476
  Flags: 0xC0 (RP--)
  Command Code: 271
  ApplicationId: 3
  Hop-by-Hop Identifier: 0x148C55A5
  End-to-End Identifier: 0xB80F145A
    {internal data}: src:gw.eap.example.net(18) rwb:(nil) rt:0 cb:(nil),(nil)
    AVP: 'Session-Id'(263) l=82 f=-M val="client.example.net;1455451008;4;cli
    AVP: 'Destination-Realm'(283) l=23 f=-M val="eap.example.net"
    AVP: 'Origin-Host'(264) l=26 f=-M val="client.example.net"
    AVP: 'Origin-Realm'(296) l=19 f=-M val="example.net"
    AVP: 'Acct-Application-Id'(259) l=12 f=-M val=3 (0x3)
    AVP: 'Acct-Session-Id'(44) l=25 f=-M val=<35 36 43 30 36 43 46 42 2D 30 3
    AVP: 'Accounting-Record-Type'(480) l=12 f=-M val='START_RECORD' (2 (0x2))
    AVP: 'Accounting-Record-Number'(485) l=12 f=-M val=3087995994 (0xb80f145a
    AVP: 'Acct-Authentic'(45) l=12 f=-M val='RADIUS' (1 (0x1))
```

```
    AVP: 'User-Name'(1) l=31 f=-M val="client1@eap.example.net"
    AVP: 'NAS-IP-Address'(4) l=12 f=-M val=<C0 A8 23 05>
    AVP: 'NAS-Identifier'(32) l=30 f=-M val="client.eap.testbed.aaa"
    AVP: 'NAS-Port'(5) l=12 f=-M val=1 (0x1)
    AVP: 'Called-Station-Id'(30) l=39 f=-M val="02-00-00-00-00-00:mac80211 te
    AVP: 'Calling-Station-Id'(31) l=25 f=-M val="02-00-00-00-01-00"
    AVP: 'NAS-Port-Type'(61) l=12 f=-M val='Wireless - IEEE 802.11 [RFC2865]'
    AVP: 'Connect-Info'(77) l=30 f=-M val="CONNECT 54Mbps 802.11g"
    AVP: 'Class'(25) l=21 f=-M val=<66 44 2F 72 67 77 78 2F 61 61 69 3A 35>

'Accounting-Answer'
  Version: 0x01
  Length: 196
  Flags: 0x40 (-P--)
  Command Code: 271
  ApplicationId: 3
  Hop-by-Hop Identifier: 0x148C55A5
  End-to-End Identifier: 0xB80F145A
    {internal data}: src:(nil)(0) rwb:(nil) rt:0 cb:(nil),(nil)((nil)) qry:0x
    AVP: 'Session-Id'(263) l=82 f=-M val="client.example.net;1455451008;4;cli
    AVP: 'Origin-Host'(264) l=30 f=-M val="server.eap.example.net"
    AVP: 'Origin-Realm'(296) l=23 f=-M val="eap.example.net"
    AVP: 'Result-Code'(268) l=12 f=-M val='DIAMETER_SUCCESS' (2001 (0x7d1))
    AVP: 'Accounting-Record-Type'(480) l=12 f=-M val='START_RECORD' (2 (0x2))
    AVP: 'Accounting-Record-Number'(485) l=12 f=-M val=3087995994 (0xb80f145a
```

6.2.9 Fault Resilience

If the primary server fails, clients are expected to fail over to alternate servers in order to minimize loss of accounting data. To guard against reboots, extended network failures, and server failures, the client should store accounting records in non-volatile memory immediately after creation and until a positive Accounting-Answer has been received from the server. The client should overwrite any previously locally stored interim accounting records to ensure that only one interim record exists for a session. Upon a reboot, the client should send stored records to the accounting server after updating the records' termination cause, session length, and other relevant information.

Diameter peers acting as agents or offline processing systems should detect duplication caused by the sending of the same record to several servers and duplication of messages in transit by inspecting the `Session-Id` and `Accounting-Record-Number` AVP pairs.

6.2.10 Example: 3GPP Rf Interface for Mobile Offline Charging

A real-world example of the Diameter base accounting application is its use in mobile networks for offline charging. Offline charging is the simultaneous collection of charging information for the use of a network resource (a voice call, data transport, or multimedia message transmission) that does not affect, in real-time, the service rendered. That is, the future use of the resource during the session is not based on the current use, for example a customer may be billed more for exceeding the number of minutes in her call plan, but she will not be cut off.

Within 3GPP networks, one of the offline charging applications is the Rf interface [5]. The Rf interface is the reference point between a 3G network element,[2] and more specifically a component within that network element known as a Charging Trigger Function (CTF), and the Charging Data Function (CDF), which collects the accounting data to send it to the network operator's billing domain.

The CTF acts as a Diameter client and the CDF acts as a stateless accounting server that implements the functionality of the Diameter base accounting application. With each chargeable event, the CTF creates an Accounting Request (ACR) and sends it to the CDF in real time. Chargeable events include the transmission of a multimedia message, the start of a session, like a voice call, a change in a session, for instance a change in quality of service, and the end of a session

The Rf interface handles both event-based and session-based accounting records. The event record is used for registration or interrogation and for successful service events (the sending of a multimedia message, for instance) triggered by a network element. In addition, event records are also used for unsuccessful session establishment attempts. The ACR record types used for accounting data related to successful sessions are start, interim, and stop. Interim events describe changes to session characteristics (e.g., tariff time switch, change of quality of service or change of session media types) or when certain limits (e.g., time or volume) are exceeded.

The CDF transfers the charging events to the network operator's billing domain via FTP.

6.2.10.1 Rf Interface Commands

The Rf interface reuses, with some modifications, the Accounting-Request and Accounting-Answer commands specified in the base protocol. The Application-Id is 3, and thus the Rf interface uses a split accounting service.

Rf Accounting-Request The Rf ACR command lacks the `Vendor-Specific-Application-Id` since the command is not part of a vendor-specific application. The command also does not use the `Accounting-Sub-Session-Id`, `Acct-Session-Id`, or `Acct-Multi-Session-Id` AVPs.

The `Accounting-Realtime-Required` AVP is not included in the request since the Rf application requires the CTF to buffer its accounting data in non-volatile memory if the CTF cannot connect to either its primary or failover CDF. Once the CTF can connect again to the CDF, it sends all stored accounting messages in the order in which they were stored.

Two new AVPs are added to Rf ACR:

`Service-Context-Id`: Contains a unique identifier of the 3GPP technical specification on which associated call detail records should be based. The AVP was originally defined in RFC 4006 [6] and expanded by 3GPP.

`Service-Information`: Holds service-specific parameters as defined by the service's charging technical specification.

---

2 3G network elements that use the Rf interface include Circuit Switched Network Elements (CS-NE), Service Network Elements (NE), SIP Application Servers (SIP AS), Media Resource Function Controllers (MRFC), Media Gateway Control Functions (MGCF), Breakout Gateway Control Functions (BGCF), Interconnect Border Control Functions (IBCF), Proxy-CSCFs (Call Session Control Function), Interrogating-CSCFs, Serving-CSCFs, Serving GPRS Support Nodes (SGSN), and Serving Gateways (S-GW)

Rf Accounting-Answer Based on the ACA defined in the base protocol, the Rf ACA command does not add any new AVPs. However, the Rf ACA command omits the following optional AVPs:

- `Vendor-Specific-Application-Id`
- `Accounting-Sub-Session-Id`
- `Acct-Session-Id`
- `Acct-Multi-Session-Id`
- `Accounting-Realtime-Required`

6.3 Credit Control

An application that monitors and controls the charges related to an end user's consumption of a service, such as network access, voice services, messaging, and downloading, is known as a credit-control application. A credit-control application checks if the user has credit available, reserves the user's credit, deducts the credit from the user's account when service is completed, and refunds any reserved credit that the user has not consumed.

The Diameter base accounting application was not designed to determine a user's account state or control the user's consumption of credit in real time. A separate application, known as the Diameter Credit-Control application and specified in [6], can be used to implement real-time credit control (Table 6.2). This generic application can be applied to many service environments that support pre-paid services.

In addition to consumption monitoring, the Diameter Credit-Control application supports service price enquiry, user's balance check, and refund of credit on the user's account. The Credit-Control application can also support a multi-service environment, where an end user can request additional service during an ongoing service, for example requesting data service during voice service.

Support for the Diameter Credit-Control application is indicated in the CER and CEA by setting the `Auth-Application-Id` to 4.

A typical credit-control architecture consists of the following actors:

Service element with an embedded Diameter credit-control client A network element, such as a NAS, SIP proxy, or an application server (gaming, messaging, etc.), that provides a service to the end users. A Diameter credit-control client monitors the usage of the granted quota according to instructions returned by the credit-control server.

Diameter credit-control server Performs real-time rating and credit control. It may also interact with a business support system. This is a logical entity, and can be combined with an AAA server.

Table 6.2 Credit-Control application.

Reference	RFC 4006 [6]	
Application-Id value	4	
Commands	Credit-Control-Request/Answer (CCR/CCA)	272

AAA server A logical entity can be combined with a credit-control server.
Business support system Provides billing support.

The credit-control client contacts the credit-control server with information about a possible service event using the Credit-Control-Request command. The credit-control application determines potential charges and verifies whether the user's account balance can cover the cost of the service being rendered and replies with a Credit-Control-Answer.

The credit-control application supports two credit authorization models. In both models, the credit-control client requests credit authorization from the credit-control server prior to allowing any service to be delivered to the end user.

Money Reservation   When the credit-control server receives a Credit-Control-Request, it reserves a suitable amount of money from the user's account and returns the corresponding amount of credit resources, which can be in money, data volume, or time units, in the Credit-Control-Answer.

Upon receipt of a successful Credit-Control-Answer, the credit-control client allows the delivery of service to the user, monitoring the user's consumption of the granted resources. When the user consumes the credit resources, or after delivery or termination of the service, the credit-control client reports the consumption to the credit-control server, which deducts the used amount from the end user's account.

Multiple requests and answers can be exchanged for session-based credit control, with the credit-control server performing rating and making new credit reservations. Both the credit-control client and the credit-control server maintain session state when engaging in session-based credit control.

Direct Debiting   Direct debiting is a single transaction, where the credit-control server deducts the money from the user's account when it receives the Credit-Control-Request. When the credit-control client receives a successful Credit-Control-Answer, it permits the service to be delivered to the end user and does not maintain session state.

6.3.1   Credit-Control-Request Command

After the user has been authenticated and authorized, the Diameter credit-control client requests credit authorization from the credit-control server for a given service with the Credit-Control-Request message (Figure 6.5).

The following AVPs occur in both the Credit-Control-Request and Credit-Control-Answer messages:

CC-Request-Type: Contains the reason for sending the message:
1   INITIAL_REQUEST: Used to initiate a credit-control session.[3]
2   UPDATE_REQUEST: Sent when the allocated quota or validity time expires and re-authorization is needed, or when service-specific events trigger an update.
3   TERMINATION_REQUEST: Sent to terminate a session.

---

3 Yes, this enumerated AVP starts with the value 1 instead of 0.

```
<Credit-Control-Request> ::= < Diameter Header: 272, REQ, PXY >
                               < Session-Id >
                               { Origin-Host }
                               { Origin-Realm }
                               { Destination-Realm }
                               { Auth-Application-Id }
                               { Service-Context-Id }
                               { CC-Request-Type }
                               { CC-Request-Number }
                               [ Destination-Host ]
                               [ User-Name ]
                               [ CC-Sub-Session-Id ]
                               [ Acct-Multi-Session-Id ]
                               [ Origin-State-Id ]
                               [ Event-Timestamp ]
                           *   [ Subscription-Id ]
                               [ Service-Identifier ]
                               [ Termination-Cause ]
                               [ Requested-Service-Unit ]
                               [ Requested-Action ]
                           *   [ Used-Service-Unit ]
                               [ Multiple-Services-Indicator ]
                           *   [ Multiple-Services-Credit-Control ]
                           *   [ Service-Parameter-Info ]
                               [ CC-Correlation-Id ]
                               [ User-Equipment-Info ]
                           *   [ Proxy-Info ]
                           *   [ Route-Record ]
                           *   [ AVP ]
```

Figure 6.5 Credit-Control-Request Command Code Format.

4 EVENT_REQUEST: Used when the credit-control server does not need to maintain session state. The request contains all information relevant to the service, and is the only request of the service. The reason for the Event request is given in the Requested-Action AVP.

CC-Request-Number: Identifies the request message within one session. The combination of Session-Id and CC-Request-Number, which is globally unique, can be used in matching credit-control requests with answers.

CC-Sub-Session-Id: Contains the credit-control sub-session identifier. Applications that require multiple credit-control sub-sessions can send messages with the same Session-Id AVP and different CC-Sub-Session-Id AVPs. If there are multiple credit sub-sessions, the sub-sessions are closed before the main session is closed to ensure proper reporting. If this AVP is not included, then there are no sub-sessions.

Multiple-Services-Credit-Control: Contains the AVPs related to the independent credit control of multiple services feature.

The following AVPs are specific to the Credit-Control-Request message:

`Service-Context-Id`: Contains a unique identifier of the service-specific document that applies to the request. Service-specific documents are documents created by standards bodies, service providers, or vendors that define the AVPs that are used as input to the rating process. The format of this AVP is a token that contains an arbitrary string, plus the domain of the standards body or the FQDN of the service provider or vendor if the identifier is allocated by a private entity and no interoperability is required.

`Subscription-Id`: Identifies the end user.

`Service-Identifier`: Contains the identifier of a service. The combination of the `Service-Identifier` and `Service-Context-Id` AVPs uniquely identifies the specific service.

`Requested-Service-Unit`: Contains the number of units requested by the credit-control client. Units can include `CC-Time` and `CC-Money`.

`Requested-Action`: Included when the `CC-Request-Type` AVP indicates an Event request, and contains the requested action:
- 0 `DIRECT_DEBITING`: Decreases the end user's account using the information in the `Requested-Service-Unit` and/or the `Service-Identifier` AVPs.
- 1 `REFUND_ACCOUNT`: Increases the end user's account using the information in the `Requested-Service-Unit` and/or the `Service-Identifier` AVPs.
- 2 `CHECK_BALANCE`: Indicates a balance check request.
- 4 `PRICE_ENQUIRY`: Indicates a price enquiry request.

`Used-Service-Unit`: Contains the number of used units measured either from the point when the service became active or when the previous measurement ended, if interim interrogations are used.

`Multiple-Services-Indicator`: Indicates whether the credit-control client can handle multiple services independently within a (sub-)session.

`Service-Parameter-Info`: Contains service-specific information used for price calculation or rating. The contents of this AVP are defined by the service-specific documentation.

`CC-Correlation-Id`: Contains information to correlate requests generated for different components of the service, for instance transport and application levels.

`User-Equipment-Info`: Indicates the identity and capabilities of the end user's terminal.

6.3.2 Credit-Control-Answer Command

The credit-control server sends a Credit-Control-Answer message (Figure 6.6) to the credit-control client to acknowledge a Credit-Control-Request command.

The following AVPs are specific to the Credit-Control-Answer message:

`CC-Session-Failover`: Indicates whether the client can move the credit-control message stream of an ongoing session to a backup server during a communication failure.

```
<Credit-Control-Answer> ::= < Diameter Header: 272, PXY >
                            < Session-Id >
                            { Result-Code }
                            { Origin-Host }
                            { Origin-Realm }
                            { Auth-Application-Id }
                            { CC-Request-Type }
                            { CC-Request-Number }
                            [ User-Name ]
                            [ CC-Session-Failover ]
                            [ CC-Sub-Session-Id ]
                            [ Acct-Multi-Session-Id ]
                            [ Origin-State-Id ]
                            [ Event-Timestamp ]
                            [ Granted-Service-Unit ]
                          * [ Multiple-Services-Credit-Control ]
                            [ Cost-Information]
                            [ Final-Unit-Indication ]
                            [ Check-Balance-Result ]
                            [ Credit-Control-Failure-Handling ]
                            [ Direct-Debiting-Failure-Handling ]
                            [ Validity-Time]
                          * [ Redirect-Host]
                            [ Redirect-Host-Usage ]
                            [ Redirect-Max-Cache-Time ]
                          * [ Proxy-Info ]
                          * [ Route-Record ]
                          * [ Failed-AVP ]
                          * [ AVP ]
```

Figure 6.6 Credit-Control-Answer Command Code Format.

Granted-Service-Unit: Contains the number of service units that the credit-control client can provide to the end user until the client must send another Credit-Control-Request.

Cost-Information: Contains cost information of a service, such as an estimate in the case of price enquiry or accumulated cost estimation in the case of a session, that the credit-control client can then provide to the end user.

Final-Unit-Indication: Indicates that the Granted-Service-Unit AVP contains the final units for the service. When these units have been exhausted, the credit-control client performs the action indicated in the Final-Unit-Action AVP given in this AVP:

0 TERMINATE: The credit-control client must terminate the service session.

1 REDIRECT: The user must be redirected, using protocols other than Diameter, to the address specified in the Redirect-Server-Address AVP. This can be a web page that allows the user to refresh his account balance, for instance.

2 RESTRICT_ACCESS: The access device must restrict the user access according to the IP packet filters defined in the Restriction-Filter-Rule AVP or or the Filter-Id AVP. Packets that do not match the filters are to be dropped.

Check-Balance-Result: Contains the result of the balance check.

Credit-Control-Failure-Handling: The credit-control client uses information in this AVP to decide what to do if sending messages to the credit-control server has been interrupted due to network issues:

0 TERMINATE: The credit-control client can only grant the service as long as it has a connection to the credit-control server. If the client does not receive a Credit-Control-Answer message within the Tx timer, then it terminates the end user's service session. This is the default behavior if the AVP is not included.

1 CONTINUE: The credit-control client should resend the request to a backup server if one exists. The client should continue to grant the service even if its credit-control messages are not delivered.

2 RETRY_AND_TERMINATE: The credit-control client should resend the request to a backup server if one is available. The client should terminate the service if its credit-control messages cannot be delivered.

Direct-Debiting-Failure-Handling: The credit-control client uses information in this AVP to decide what to do if sending direct-debiting messages to the credit-control server has been interrupted due to network issues:

0 TERMINATE_OR_BUFFER: The credit-control client grants the service as long as there is a connection to the credit-control server. If the client does not receive a Credit-Control-Answer message within the Tx timer, then the client terminates the service if it can determine from the failed answer that units have not been debited. Otherwise the client grants the service and tries to re-send the request. This is the default behavior if the AVP is not included.

1 CONTINUE: The client should continue to grant the service even if its credit-control messages are not delivered.

Validity-Time: Contains the validity time of the granted service units. The credit-control client starts to measure the the validity time when it receives the Credit-Control-Answer. If the granted service units have not been consumed within the specified validity time, the credit-control client sends a Credit-Control-Request with CC-Request-Type set to UPDATE_REQUEST.

6.3.3 Failure Handling

To ensure that the end user's account is not debited or credited multiple times for the same service event, only one place in the credit-control system should perform duplicate detection.

As covered in Section 6.3.2, the credit-control server provides guidance to the credit-control client on how to handle communication interruptions with two optional AVPs:

- Credit-Control-Failure-Handling
- Direct-Debiting-Failure-Handling

If the Credit-Control-Answer does not contain the above AVPs, then the credit-control client grants service as long as it has a connection to the credit-control server, otherwise it terminates the end user's service session.

6.3.4 Extensibility

Because the credit-control application is a generic framework, it has been made extensible, and use of service-specific documents to define inputs to the rating process is expected. It is assumed that any service-specific AVPs are known to both the credit-control client and the credit-control server before Diameter credit-control messages are exchanged between them, and the document that defines those AVPs is captured in the `Service-Context-Id` AVP of the Credit-Control-Request. To help with interoperability, it is preferable that new credit-control AVPs are defined within a standards organization. For example, 3GPP has several service-specific documents, and the `Service-Context-Id` AVP contains a reference to the technical specification that defines the AVPs.

At the time of writing of this book, RFC 4006 is being updated by the IETF. Changes include updating references, fixing errata, improving the security section, clarifying details of some of the AVPs (`Used-Service-Unit`, `Redirect-Address-Type`, `Filter-Rule`), and adding a few new ones to extend current AVPs, such as `User-Equipment-Info-Extension`, `Subscription-Id-Extension`, `Redirect-Server-Extension`, and `QoS-Final-Unit-Indication`, which supports quality of service.

6.3.5 Example: 3GPP Ro Interface for Online Charging

3GPP uses a credit-control application on the Ro interface for online charging for SIP services, the details of which are captured in TS 32.240 [7] and TS 32.299 [5]. The Ro interface supports real-time transactions, and both stateless (event-based charging) and stateful (session-based charging) modes of operation. The Ro interface exists between a charging trigger function (CTF or credit-control client) and an online charging system (OCS or credit-control server). The CTF is a 3G network element and can be a media resource function controller or a SIP application server. The CTF sends charging events via Credit-Control-Requests for online charging to the OCS.

Two new AVPs are added to the Credit-Control-Request command:

`AoC-Request-Type`: Denotes if the user equipment (UE) has requested advice of charge (AoC) and what type of information, such as cost or tariff information, is needed.

`Service-Information`: Holds service-specific parameters as defined by the service's charging technical specification.

The following AVPs defined for the Credit-Control-Request command are not used in 3GPP networks:

- `CC-Sub-Session-Id`
- `Acct-Multi-Session-Id`
- `Service-Identifier`
- `Requested-Service-Unit`
- `Used-Service-Unit`
- `Service-Parameter-Info`

The OCS sends a Credit-Control-Answer to acknowledge these charging events to the CTF to grant or reject the network resource usage requested in the charging event, according to the decision taken by the online charging system.

The following AVPs defined for the Credit-Control-Answer command are not used in 3GPP networks:

- `User-Name`
- `CC-Sub-Session-Id`
- `Acct-Multi-Session-Id`
- `Origin-State-Id`
- `Event-Timestamp`
- `Granted-Service-Unit`
- `Final-Unit-Indication`
- `Check-Balance-Result`
- `Validity-Time`

Although the protocol may be stateless for event-based charging, the CTF and OCS may keep internal state across the different charging events.

6.4   Quality of Service

Quality of Service (QoS) refers to the performance of a network from the user's perspective. To provide a certain level of QoS, the network operator takes performance measurements and acts upon them. Operators may provide different levels of QoS to different users if users have paid for a greater or lesser level of service.

RFC 5866 defines the framework and messages used for the Diameter QoS application. The commands specific to the QoS application are given in Table 6.3.

The QoS application also uses the Re-Auth-Request/Answer (RAR/RAA), Abort-Session-Request/Answer (ASR/ASA), and Session-Term-Request/Answer (STR/STA) commands from the base protocol.

6.4.1   Actors

The actors in the Diameter QoS application include both Diameter and non-Diameter entities. Although RFC 5866 assigns stylized acronyms to all of the actors, it does not rigidly define the behavior of the non-Diameter actors:

application endpoint (AppE) This is the end user's signaling application. The AppE exchanges messages, for instance SIP messages, with an application server and/or with other AppEs. It may or may not have the ability to request QoS enhancement for its signaling messages.

Table 6.3 Quality of service application.

References	RFC 5866 [8], RFC 5777 [9], RFC 5624 [10]	
Application-Id value	9	
Commands	QoS-Authorization-Request/Answer (QAR/QAA)	326
	QoS-Install-Request/Answer (QIR/QIA)	327

application server This non-Diameter node supports the signaling protocol. Depending on the signaling protocol used, the application server may be a source of authorization for QoS-enhanced application flows.

network element (NE) This is the Diameter client in the QoS application and is a QoS-aware router. It can effect QoS changes when it is given information from the Diameter server (AE) that handles the Diameter QoS application.

authorizing entity (AE) This is the Diameter server in the QoS application and authorizes QoS requests. It may be a standalone entity or may be integrated with an application server.

resource-requesting entity (RRE) This non-Diameter, logical node supports the signaling protocol interaction for QoS resources. The RRE resides in the AppE and communicates with its peer logical entity in the AE or NE to trigger the QoS authorization process. The communication between the RRE and the NE or AE is not defined by the QoS application.

The Diameter QoS application is executed between the NE and AE.

6.4.2 Modes of Operation

The Diameter QoS mechanisms can be divided into two modes: Push or Pull. Which mode is used depends on the AppE's ability to initiate QoS resource requests. If the AppE cannot initiate QoS requests, Push mode is used. If the AppE can initiate QoS requests, Pull mode is used.

After receiving the authorization request from the application server or the NE, the AE decides the appropriate mode (Push or Pull) based on information received from the request and/or based on configuration.

6.4.2.1 Push Mode

In the Push mode, since the AppE cannot request the QoS resource authorization on its own, the authorization process is triggered either by the application server's request or by policies defined by the operator. For example, an operator policy may be to modify QoS depending on time of day in order to improve throughput.

The AE authenticates and authorizes the QoS request based on information from the application server. The AE then determines to which NE(s) to push the QoS authorization information, with the selection also being based on information received from the application server, such as the IP addresses of media flows. The AE then sends a QoS-Install-Request message to the NE to apply the QoS authorization state. The NE acknowledges the QoS instruction with a QoS-Install-Answer message.

6.4.2.2 Pull Mode

Pull mode provides QoS for AppEs and NEs that have the ability to request QoS authorization. The authorization process is initiated when NE receives the network signaling from the AppE or a local event occurs in the NE according to pre-configured policies. The NE requests QoS authorization from the AE with a QoS-Authorization-Request command. The AE returns the result of the authorization decision in a QoS-Authorization-Answer command.

6.4.3 Authorization

6.4.3.1 Push Mode Authorization Schemes

Push mode can be divided into two types: endpoint-initiated and network-initiated.

In the endpoint-initiated mode (Figure 6.7), the RRE sends a QoS request to the application server via application signaling. The application server extracts application-level QoS information and triggers the authorization process to the AE.

In the network-initiated mode, the application server triggers the authorization process without a QoS request from the endpoint.

If the endpoint does not indicate QoS attributes, the AE and/or application server determines the QoS requirements from a combination of application-level QoS information, network policies, end-user subscription, and network resource availability.

The AE makes an authorization decision and sends the decision to the NE directly.

6.4.3.2 Pull Mode Authorization

Authorization schemes for Pull mode can involve either two or three parties. From the Diameter QoS application's point of view, these authorization schemes differ only in the type of information carried.

In the two-party scheme, the NE authenticates the QoS RRE. The NE either makes authorization decisions locally or delegates the decision to a trusted entity. No Diameter QoS protocol interaction is required for this authorization scheme.

In the basic three-party scheme (Figure 6.8), the NE forwards a received QoS reservation request to the AE, which makes the authorization decision.

The figure also shows an accounting step to track resources used by the application endpoint.

The token-based three-party scheme (Figure 6.9) uses authorization tokens to assist the authorization process at the NE or the AE. Tokens can assist in preventing fraud and ensuring accurate billing. RFC 3521 [11], Framework for Session Set-up with Media Authorization, covers using tokens in establishing media streams with QoS.

The RRE requests an authorization token from the AE via the application layer. If the authorization decision is successful, the AE generates an authorization token and sends it to the RRE for inclusion in the QoS signaling protocol. For an RRE/AE interaction

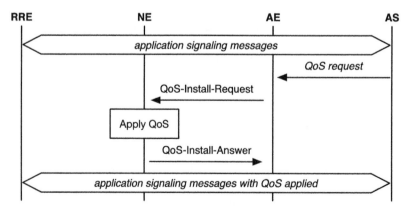

Figure 6.7 Authorization scheme for Push mode.

Figure 6.8 Three-party authorization scheme for Pull mode.

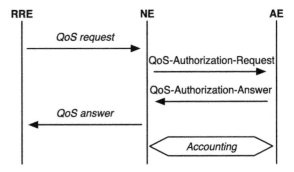

Figure 6.9 Token-based three-party scheme for Pull mode.

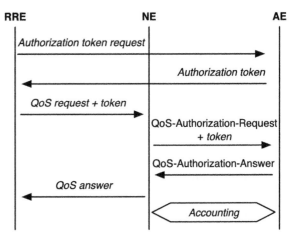

example, refer to RFC 3313 [12], which specifies Session Initiation Protocol (SIP) extensions for media authorization.

The authorization token associates the application-layer authorization decision with the corresponding QoS signaling session. At a minimum the authorization token contains the following authorization parameters:

- an identifier for the issuing AE,
- an identifier for the application protocol session for which the token was issued,
- a keyed message digest or digital signature protecting the token's content.

An example structure for the authorization token can be found in RFC 3520 [13], Session Authorization Policy Element.

The RRE sends the QoS signaling message containing the authorization token to the NE. The NE uses the authorization token to authorize the QoS request. Alternatively, the NE can forward the authorization token to the AE in the user's home network using the Diameter QoS application.

In scenarios where the RRE interacts with the AE at the application layer, binding mechanisms that do not use tokens can also work. For instance, the application-layer protocol interaction may inform the RRE of the transport port numbers where it will receive media streams. The RRE then informs the NE about the port numbers in an IP filter indication so that the NE can tunnel the inbound packets. The NE forwards the RRE's IP address and the IP filter indications to the AE in the QoS authorization request. The AE matches the filter information with the port numbers given in the earlier application-layer protocol interaction and identifies the policy information to send to the NE in its QoS authorization response.

6.4.4 Establishing and Managing a QoS Application Session

6.4.4.1 Establishing a Session

Pull and Push modes use different Diameter messages for session establishment. Pull mode uses QoS-Authorization-Request/Answer, and Push mode uses QoS-Install-Request/Answer. However, they use the same set of base commands for other operations, such as session modification (RAR/RAA) and termination (STR/STA).

The QoS authorization session is typically established on a per-subscriber basis (i.e., all requests with the same User-ID), but it can also be established on a per-node or per-request basis. The concurrent sessions between an NE and an AE are identified by different Session-Ids.

Session Establishment for Pull Mode A Diameter QoS authorization session starts when the NE receives a request for a QoS reservation or local events trigger it. The NE converts the following information from the QoS signaling message to Diameter AVPs and generates a QAR message (Figure 6.10):

- The AE's identity goes into the `Destination-Host` and `Destination-Realm` AVPs.
- The RRE's identity is used to create the `User-Name` AVP.
- The `QoS-Authorization-Data` AVP contains the RRE's authorization token credentials.
- Requested QoS parameters are carried in the `QoS-Resources` AVP, which is defined in RFC 5777 [9].

The NE also includes the following information in the QAR message:

- information about the application session,
- the signaling session identifier and/or QoS-enabled data flows identifier.

```
<QoS-Authorization-Request> ::= < Diameter Header: 326, REQ, PXY >
                                < Session-Id >
                                { Auth-Application-Id }
                                { Origin-Host }
                                { Origin-Realm }
                                { Destination-Realm }
                                { Auth-Request-Type }
                                [ Destination-Host ]
                                [ User-Name ]
                            *   [ QoS-Resources ]
                                [ QoS-Authorization-Data ]
                                [ Bound-Auth-Session-Id ]
                            *   [ AVP ]
```

Figure 6.10 QoS-Authorization-Request Command Code Format.

```
<QoS-Authorization-Answer> ::= < Diameter Header: 326, PXY >
                                < Session-Id >
                                { Auth-Application-Id }
                                { Auth-Request-Type }
                                { Result-Code }
                                { Origin-Host }
                                { Origin-Realm }
                            *   [ QoS-Resources ]
                                [ Acct-Multisession-Id ]
                                [ Session-Timeout ]
                                [ Authorization-Session-Lifetime ]
                                [ Authorization-Grace-Period ]
                            *   [ AVP ]
```

Figure 6.11 QoS-Authorization-Answer Command Code Format.

The NE sends the QAR to an AE. When the AE receives the QAR, it starts authorization processing. Based on the information in the QoS-Authorization-Data, User-Name, and QoS-Resources AVPs, the AE determines the QoS resources and flow state (enabled/disabled) from locally available information (e.g., application-layer policy or the user's subscription profile). The AE sends its decision to the NE in its QoS-Authorization-Answer (QAA) message (Figure 6.11).

The AE saves authorization session state and session state information (e.g., Signaling-Session-Id, authentication data) for management of the session.

The Result-Code AVP of the QAA message contains the result of the authorization request. If the authorization was successful (DIAMETER_LIMITED_SUCCESS), the QoS-Resources AVP will contain information about the authorized QoS resources and the status of the authorized flow (enabled/disabled). The QoS information provided via the QAA is applied by the QoS Traffic Control function of the NE.

The value DIAMETER_LIMITED_SUCCESS indicates that the AE expects confirmation via another QAR message for successful QoS resource reservation and for final reserved QoS resources.

The QAA also contains the Authorization-Session-Lifetime AVP, which states the length of time for which the authorization decision is valid and is determined by the user's subscription profile and/or credit availability. To extend the authorization period, the NE will need to send a new QoS-Authorization-Request.

The NE sends another QAR to indicate successful QoS reservation and activation of the data flow. This QAR contains the established QoS resources, which may be different than those authorized with the initial QAA due to QoS-signaling-specific behavior or the result of QoS negotiation along the data path. The AE acknowledges the reserved QoS resources by setting the Result-Code AVP to DIAMETER_SUCCESS in its QAA.

Session Establishment for Push Mode The AE starts authorization processing when it receives a request from an RRE through an application server or when it is triggered by a local event. Based on information in the request, the AE determines the authorized QoS resources and flow state (enabled/disabled) from locally available information (e.g., policy information that may be previously established as part of an application-layer signaling exchange, or the user's subscription profile). The AE determines how to

```
<QoS-Install-Request> ::= < Diameter Header: 327, REQ, PXY >
                         < Session-Id >
                         { Auth-Application-Id }
                         { Origin-Host }
                         { Origin-Realm }
                         { Destination-Realm }
                         { Auth-Request-Type }
                         [ Destination-Host ]
                       * [ QoS-Resources ]
                         [ Session-Timeout ]
                         [ Authorization-Session-Lifetime ]
                         [ Authorization-Grace-Period ]
                         [ Authorization-Session-Volume ]
                       * [ AVP ]
```

Figure 6.12 QoS-Install-Request Command Code Format.

contact the NE via base protocol peer discovery mechanisms and determines the data flow for which the QoS reservation should be established. Then the AE sends its authorization decision in the QoS-Install-Request (QIR) to the NE (Figure 6.12).

The following information is contained in the QIR:

- The NE's identity goes into the `Destination-Realm` and `Destination-Host` AVPs.
- The RRE's identity is used to populate the `User-Name` AVP.
- The `QoS-Authorization-Data` AVP contains the RRE's authorization token credentials.
- Requested QoS parameters are used to create the `QoS-Resources` AVP, which is defined in RFC 5777 [9].
- Information about the application session.
- The signaling session identifier and/or QoS-enabled data flows identifier.

The AE keeps authorization session state and information for management of the session (e.g., `Signaling-Session-Id`, authentication data).

The AE provides the authorization decision in the `QoS-Resources` AVP of the QIR. The QoS information provided in the QIR is applied by the QoS Traffic Control function of the NE.

The `Authorization-Session-Lifetime` AVP allows the NE to determine the validity period of the QoS reservation. The user's subscriber profile and account standing may influence the session's duration. To extend the authorization period, the AE should send a new QoS-Install-Request.

The NE sends QoS-Install-Answer (Figure 6.13) to indicate QoS reservation and activation of the signaling data flow. To indicate success, the NE sets the `Result-Code` AVP to `DIAMETER_SUCCESS`. The QIA contains the established QoS resources, which may be different than those authorized with the initial QIR due to QoS-signaling-specific behavior or the result of QoS negotiation along the data path.

```
<QoS-Install-Answer> ::= < Diameter Header: 327, PXY >
                         < Session-Id >
                         { Auth-Application-Id }
                         { Origin-Host }
                         { Origin-Realm }
                         { Result-Code }
                       * [ QoS-Resources ]
                       * [ AVP ]
```

Figure 6.13 QoS-Install-Answer Command Code Format

6.4.5 Re-Authorizing a Session

Re-authorization policies depend on the QoS signaling protocol.

6.4.5.1 Re-Authorization Initiated by the NE
In the `Authorization-Session-Lifetime` AVP of the QAA, the AE specifies the duration of the authorization session. The NE may send a new QAR when the NE receives a QoS signaling message that requires modification of the authorized parameters of an ongoing QoS session or the authorization lifetime expiration.

6.4.5.2 Re-Authorization Initiated by the Authorizing Elements
The AE may use a base protocol Re-Authorization-Request (RAR) to perform re-authorization with the authorized parameters directly when the re-authorization is triggered by a service request, local events, or policy rules. The RAR may include the parameters of the re-authorized QoS state such as reserved resources, the duration of the reservation, and identification of the QoS-enabled flow or signaling session.

When the NE receives a RAR message, it checks to see that the Session-Id matches an active QoS session. The NE then sends a Re-Auth-Answer (RAA) message to the AE.

If the RAR does not include any parameters of the re-authorized QoS state, the NE sends a QoS-Authorization-Request (QAR) message to the AE.

6.4.6 Terminating a Session

6.4.6.1 Session Terminated by the NE
A session can be terminated due to the expiration of the QoS reservation soft state or due to a request via a QoS signaling message to delete the QoS reservation state.

To end the authorization session for an installed QoS reservation state, the NE sends a Session-Termination-Request (STR) message to the AE. The AE replies with a Session-Termination-Acknowledgement (STA) message.

6.4.6.2 Session Terminated by the AE
An AE may terminate a session due to insufficient credits or session termination at the application layer.

To end a session, the AE sends an Abort-Session-Request (ASR) message to the NE. The NE notifies the application endpoint and deletes the QoS reservation state. The NE then sends an Abort-Session-Answer (ASA) message to the AE, which deletes the authorization session state it had been maintaining.

6.5 Interworking RADIUS and Diameter

Diameter was designed to allow transitioning from RADIUS to Diameter in existing deployments, where it may not be realistic to replace all the network components at once. To address this need, Diameter defined a type of gateway, called a *translation agent*, to convert from RADIUS to Diameter. The RFC 4005 (*Diameter Network Access Server Application*) specification [14] documented basic guidelines for RADIUS/Diameter protocol interactions. In addition, many Diameter protocol values (e.g., AVP codes) have been aligned with RADIUS protocol values to simplify the conversion work. Furthermore, the IETF required each RADIUS RFC to include a *Diameter Considerations* section to describe how translation between the two protocols could be achieved. It was rather quickly noticed that translation between RADIUS and Diameter only works for very simple AAA interactions or for interactions that are specified in fine detail for each use case/application (i.e., the behavior of the translation agent). The RFC (RFC 7155 [2]) that obsoleted RFC4005 removed all material regarding RADIUS/Diameter protocol interactions, and the IETF no longer requires *Diameter Considerations* sections in RADIUS specifications. Nowadays, the gap between RADIUS and Diameter has diminished such that only the concepts of the application and capabilities exchange are missing from RADIUS when compared to Diameter. It has been accepted that these two protocols evolve separately.

There is, however, no magic here. Converting between the protocols, accommodating different state machines, and a few fundamental differences between the protocol designs make the implementation of a conversion gateway both difficult and limited to specific Diameter applications in general.

freeDiameter comes with a specific example of such gateway (i.e., translation agent functionality) to convert between a RADIUS client application and a Diameter backend server. The gateway functionality is limited to the Diameter EAP application (Table 6.4) and the Diameter Base Accounting application, but it can serve as a basis to extend the implementation when useful.

In order to set up the test environment for deploying the gateway by yourself, you will need to create three clones of the freeDiameter VM; please refer to Appendix A for information on this VM if you have not already. The three clones to be created are the following:

eap.srv The backend server that runs the Diameter EAP application backend as well as the Diameter Accounting server.

Table 6.4 Extensible Authentication Protocol (EAP) application.

Reference	RFC 4072 [15]
Application-Id value	5
Commands	Diameter-EAP-Request/Answer (DER/DEA) 268

eap.gw The conversion gateway between RADIUS and Diameter.

eap.cli The RADIUS client application.

After you have created the VMs listed above, here are the configuration steps to run on each VM:

eap.srv The backend is preconfigured for you; just execute the following commands in a Terminal window to get it started: `$ nw_configure.sh server.eap` `.example.net` to set up the network configuration, `$ fD_configure.sh` `7_srv` to compile and configure the relevant `freeDiameter` extensions, and finally `$ freeDiameterd` to start the server.

eap.gw The gateway follows the same steps: `$ nw_configure.sh gw.eap.` `example.net`, `$ fD_configure.sh 7_gw`, and `$ freeDiameterd`. If the server is already running, you should see both instances connecting to each others at this point.

eap.cli This machine contains the RADIUS client application. We are using **hostapd**, a software access point implementation, as the EAP authenticator and RADIUS client, and **wpasupplicant** as the EAP supplicant. They communicate over a virtual WiFi interface emulated by the Linux kernel. The steps to configure this machine are slightly different: `$ nw_configure.sh client.eap.example.net`, `$ fD_configure.sh 7_cli`, then we are not starting **freeDiameterd** since this machine is a RADIUS client. Instead, two commands are available to help triggering an authentication. Open two Terminal windows, in the first one run `$ run-hostapd.sh`. In the other Terminal window, run `$ run-wpasupplicant.sh`.

When the backend server starts, it loads two important extensions. *app_acct.fdx* is a Diameter Accounting server implementation that establishes a connection to a local PostgreSQL database in which it stores the accounting records it receives. *app_diameap.fdx* is a Diameter EAP backend application that authenticates clients based on a configuration stored in a local MySQL database. The environment is pre-configured so that a user "client1" is authenticated using EAP-TLS.

When the gateway starts, its Diameter side connects to the backend server as a client. At the same time the gateway opens RADIUS server ports for authentication and accounting, and waits for incoming connections. No additional activity happens until messages are received, except for the regular watchdog exchange with the server.

On the client, when **hostapd** starts (which is the typical implementation used in many WiFi access points) its configuration tells it to connect to the RADIUS backend located on the gateway. An initial RADIUS accounting message is sent to the gateway that replies (without converting to Diameter in that case, as this is an example of behavior that does not translate to Diameter) and **hostapd** is ready to handle incoming clients.

Finally, when **wpasupplicant** is started, it emulates a client connecting to the access point. This client is configured to use the "client1" identity with a specific private key and certificate that matches the configuration of the backend server.

As a first step, an EAP exchange takes place between the client and the backend server. The exchange is carried over emulated WPA between **wpasupplicant** and **hostapd**, then over RADIUS between **hostapd** and the gateway, and finally over Diameter between the gateway and the backend server. Responses follow the same reverse path. The exchange goes on until the backend server is able to authenticate the client successfully and derive key material to secure the WPA link. This key material is sent to **hostapd** through an EAP-Master-Session-Key AVP that gets converted to its RADIUS equivalent by the gateway. On the other side, the client has derived the same key material during the EAP exchange. The WPA link is now secured with key derived from this material.

Once the authentication and authorization are successful, **hostapd** sends accounting records for the session in progress. The gateway receives these and converts them to Diameter as well, using the Diameter Accounting application this time. The resulting messages are handled by the appropriate server extension on the backend. In a real-life deployment case, the accounting server would probably be a different physical entity from the authentication server, but this is not important for the purpose of this example. As long as the client remains connected to the **hostapd** instance, the accounting session continues. Depending on the configuration, interim records may be exchanged (this can be forced in *hostapd.conf* file using the "radius_acct_interim_interval=60" parameter, for example). Finally, when the client disconnects, which can be emulated by stopping the **wpasupplicant** instance, **hostapd** sends a final accounting record for the session and cleans its state machine.

Here is an edited log from the backend server showing the process.

```
# wpasupplicant is started, the EAP exchange starts
RCV from 'gw.eap.example.net':
     'Diameter-EAP-Request'
        Version: 0x01
        Length: 460
        Flags: 0xC0 (RP--)
        Command Code: 268
        ApplicationId: 5
        Hop-by-Hop Identifier: 0x148C55A0
        End-to-End Identifier: 0xB80F1455
        AVP: 'Session-Id'(263) l=82 f=-M
     val="client.example.net;1455451008;4;client1@eap.example.net
     ;gw.eap.example.net"
# Above Session-Id value is generated by the Gateway
        AVP: 'Destination-Realm'(283) l=23 f=-M val="eap.example.net"
        AVP: 'Origin-Host'(264) l=26 f=-M val="client.example.net"
        AVP: 'Origin-Realm'(296) l=19 f=-M val="example.net"
        AVP: 'Auth-Application-Id'(258) l=12 f=-M val=5 (0x5)
        AVP: 'Auth-Request-Type'(274) l=12 f=-M val='
     AUTHORIZE_AUTHENTICATE'
        AVP: 'Origin-AAA-Protocol'(408) l=12 f=-M val='RADIUS' (1 (0x1)
     )
# Above AVP indicates that this is a translated AAA session
        AVP: 'EAP-Payload'(462) l=36 f=-M val=<02 C2 00 1C 01 63 6C 69
     65 6E
        AVP: 'User-Name'(1) l=31 f=-M val="client1@eap.example.net"
        AVP: 'NAS-IP-Address'(4) l=12 f=-M val=<C0 A8 23 05>
        AVP: 'NAS-Identifier'(32) l=30 f=-M val="client.eap.testbed.aaa
     "
        AVP: 'NAS-Port'(5) l=12 f=-M val=1 (0x1)
        AVP: 'Called-Station-Id'(30) l=39 f=-M val="02-00-00-00-00-00:
     mac802
        AVP: 'Calling-Station-Id'(31) l=25 f=-M val="02-00-00-00-01-00"
```

```
        AVP: 'Framed-MTU'(12) l=12 f=-M val=1400 (0x578)
        AVP: 'NAS-Port-Type'(61) l=12 f=-M val='Wireless - IEEE 802.11
    [RFC2
        AVP: 'Connect-Info'(77) l=30 f=-M val="CONNECT 54Mbps 802.11g"
SND to 'gw.eap.example.net':
    'Diameter-EAP-Answer'
        Version: 0x01
        Length: 256
        Flags: 0x40 (-P--)
        Command Code: 268
        ApplicationId: 5
        Hop-by-Hop Identifier: 0x148C55A0
        End-to-End Identifier: 0xB80F1455
        AVP: 'Session-Id'(263) l=82 f=-M val="client.example.net
    ;1455451008;4;client1@eap.example.net;gw.eap.example.net"
        AVP: 'Origin-Host'(264) l=30 f=-M val="server.eap.example.net"
        AVP: 'Origin-Realm'(296) l=23 f=-M val="eap.example.net"
        AVP: 'Auth-Application-Id'(258) l=12 f=-M val=5 (0x5)
        AVP: 'Auth-Request-Type'(274) l=12 f=-M val='AUTHENTICATE_ONLY'
    (1
        AVP: 'User-Name'(1) l=31 f=-M val="client1@eap.example.net"
        AVP: 'Result-Code'(268) l=12 f=-M val='
    DIAMETER_MULTI_ROUND_AUTH'
# Above AVP indicates that another EAP exchange at least is needed
        AVP: 'Multi-Round-Time-Out'(272) l=12 f=-M val=30 (0x1e)
        AVP: 'EAP-Payload'(462) l=14 f=-M val=<01 C3 00 06 0D 20>


# the exchange continues for several rounds
RCV from 'gw.eap.example.net':
    'Diameter-EAP-Request'
SND to 'gw.eap.example.net':
    'Diameter-EAP-Answer'
RCV from 'gw.eap.example.net':
    'Diameter-EAP-Request'
SND to 'gw.eap.example.net':
    'Diameter-EAP-Answer'
RCV from 'gw.eap.example.net':
    'Diameter-EAP-Request'
SND to 'gw.eap.example.net':
    'Diameter-EAP-Answer'


RCV from 'gw.eap.example.net':
    'Diameter-EAP-Request'
        Version: 0x01
        Length: 472
        Flags: 0xC0 (RP--)
        Command Code: 268
        ApplicationId: 5
        Hop-by-Hop Identifier: 0x148C55A4
        End-to-End Identifier: 0xB80F1459
        AVP: 'Session-Id'(263) l=82 f=-M val="client.example.net
    ;1455451008;4;client1@eap.example.net;gw.eap.example.net"
        AVP: 'Destination-Host'(293) l=30 f=-M val="server.eap.example.
    net"
        AVP: 'Destination-Realm'(283) l=23 f=-M val="eap.example.net"
        AVP: 'Origin-Host'(264) l=26 f=-M val="client.example.net"
        AVP: 'Origin-Realm'(296) l=19 f=-M val="example.net"
        AVP: 'Auth-Application-Id'(258) l=12 f=-M val=5 (0x5)
```

```
        AVP: 'Auth-Request-Type'(274) l=12 f=-M val='
    AUTHORIZE_AUTHENTICATE'
        AVP: 'Origin-AAA-Protocol'(408) l=12 f=-M val='RADIUS' (1 (0x1)
    )
        AVP: 'EAP-Payload'(462) l=14 f=-M val=<02 C6 00 06 0D 00>
        AVP: 'User-Name'(1) l=31 f=-M val="client1@eap.example.net
        AVP: 'NAS-IP-Address'(4) l=12 f=-M val=<C0 A8 23 05>
        AVP: 'NAS-Identifier'(32) l=30 f=-M val="client.eap.testbed.aaa
    "
        AVP: 'NAS-Port'(5) l=12 f=-M val=1 (0x1)
        AVP: 'Called-Station-Id'(30) l=39 f=-M val="02-00-00-00-00-00:
    mac802
        AVP: 'Calling-Station-Id'(31) l=25 f=-M val="02-00-00-00-01-00"
        AVP: 'Framed-MTU'(12) l=12 f=-M val=1400 (0x578)
        AVP: 'NAS-Port-Type'(61) l=12 f=-M val='Wireless - IEEE 802.11
    [RFC2
        AVP: 'Connect-Info'(77) l=30 f=-M val="CONNECT 54Mbps 802.11g"

[DiamEAP extension]  Auth Success: client1@eap.example.net
# app_diameap indicates a successful authentication

SND to 'gw.eap.example.net':
    'Diameter-EAP-Answer'
      Version: 0x01
      Length: 328
      Flags: 0x40 (-P--)
      Command Code: 268
      ApplicationId: 5
      Hop-by-Hop Identifier: 0x148C55A4
      End-to-End Identifier: 0xB80F1459
      AVP: 'Session-Id'(263) l=82 f=-M val="client.example.net
    ;1455451008;4;client1@eap.example.net;gw.eap.example.net"
      AVP: 'Origin-Host'(264) l=30 f=-M val="server.eap.example.net"
      AVP: 'Origin-Realm'(296) l=23 f=-M val="eap.example.net"
      AVP: 'Auth-Application-Id'(258) l=12 f=-M val=5 (0x5)
      AVP: 'Auth-Request-Type'(274) l=12 f=-M val='AUTHENTICATE_ONLY'
    (1
      AVP: 'User-Name'(1) l=31 f=-M val="client1@eap.example.net"
      AVP: 'Result-Code'(268) l=12 f=-M val='DIAMETER_SUCCESS' (2001
    (0x7
      AVP: 'EAP-Payload'(462) l=12 f=-M val=<03 C7 00 04>
# Here is the key material for hostapd
      AVP: 'EAP-Master-Session-Key'(464) l=72 f=-M val=<9F A3 EE D0
    B0...
      AVP: 'Accounting-EAP-Auth-Method'(465) l=16 f=-M val=13 (0xd)

# Authentication is done, now starts accounting
RCV from 'gw.eap.example.net':
    'Accounting-Request'
      Version: 0x01
      Length: 476
      Flags: 0xC0 (RP--)
      Command Code: 271
      ApplicationId: 3
      Hop-by-Hop Identifier: 0x148C55A5
      End-to-End Identifier: 0xB80F145A
      AVP: 'Session-Id'(263) l=82 f=-M val="client.example.net
    ;1455451008;4;client1@eap.example.net;gw.eap.example.net"
      AVP: 'Destination-Realm'(283) l=23 f=-M val="eap.example.net"
```

```
    AVP: 'Origin-Host'(264) l=26 f=-M val="client.example.net"
    AVP: 'Origin-Realm'(296) l=19 f=-M val="example.net"
    AVP: 'Acct-Application-Id'(259) l=12 f=-M val=3 (0x3)
    AVP: 'Acct-Session-Id'(44) l=25 f=-M val=<35 36 43 30 36 43 46
42 2D
    AVP: 'Accounting-Record-Type'(480) l=12 f=-M val='START_RECORD'
(2
    AVP: 'Accounting-Record-Number'(485) l=12 f=-M val=3087995994
(0xb80
    AVP: 'Acct-Authentic'(45) l=12 f=-M val='RADIUS' (1 (0x1))
    AVP: 'User-Name'(1) l=31 f=-M val="client1@eap.example.net"
    AVP: 'NAS-IP-Address'(4) l=12 f=-M val=<C0 A8 23 05>
    AVP: 'NAS-Identifier'(32) l=30 f=-M val="client.eap.testbed.aaa
"
    AVP: 'NAS-Port'(5) l=12 f=-M val=1 (0x1)
    AVP: 'Called-Station-Id'(30) l=39 f=-M val="02-00-00-00-00-00:
mac802
    AVP: 'Calling-Station-Id'(31) l=25 f=-M val="02-00-00-00-01-00"
    AVP: 'NAS-Port-Type'(61) l=12 f=-M val='Wireless - IEEE 802.11
[RFC2
    AVP: 'Connect-Info'(77) l=30 f=-M val="CONNECT 54Mbps 802.11g"
    AVP: 'Class'(25) l=21 f=-M val=<66 44 2F 72 67 77 78 2F 61 61
69 3A
SND to 'gw.eap.example.net':
    'Accounting-Answer'
    Version: 0x01
    Length: 196
    Flags: 0x40 (-P--)
    Command Code: 271
    ApplicationId: 3
    Hop-by-Hop Identifier: 0x148C55A5
    End-to-End Identifier: 0xB80F145A
    AVP: 'Session-Id'(263) l=82 f=-M val="client.example.net
;1455451008;4;client1@eap.example.net;gw.eap.example.net"
    AVP: 'Origin-Host'(264) l=30 f=-M val="server.eap.example.net"
    AVP: 'Origin-Realm'(296) l=23 f=-M val="eap.example.net"
    AVP: 'Result-Code'(268) l=12 f=-M val='DIAMETER_SUCCESS' (2001
(0x7d
    AVP: 'Accounting-Record-Type'(480) l=12 f=-M val='START_RECORD'
(2 (
    AVP: 'Accounting-Record-Number'(485) l=12 f=-M val=3087995994
(0xb80

# The client disconnects
RCV from 'gw.eap.example.net':
    'Accounting-Request'
    Version: 0x01
    Length: 564
    Flags: 0xC0 (RP--)
    Command Code: 271
    ApplicationId: 3
    Hop-by-Hop Identifier: 0x148C55AE
    End-to-End Identifier: 0xB80F1463
    AVP: 'Session-Id'(263) l=82 f=-M val="client.example.net
;1455451008;4;client1@eap.example.net;gw.eap.example.net"
    AVP: 'Destination-Realm'(283) l=23 f=-M val="eap.example.net"
    AVP: 'Origin-Host'(264) l=26 f=-M val="client.example.net"
    AVP: 'Origin-Realm'(296) l=19 f=-M val="example.net"
    AVP: 'Acct-Application-Id'(259) l=12 f=-M val=3 (0x3)
```

```
        AVP: 'Acct-Session-Id'(44) l=25 f=-M val=<35 36 43 30 36 43 46 42
    2D
        AVP: 'Accounting-Record-Type'(480) l=12 f=-M val='STOP_RECORD'
    (4 (0
        AVP: 'Accounting-Record-Number'(485) l=12 f=-M val=3087996003
    (0xb80
        AVP: 'Acct-Authentic'(45) l=12 f=-M val='RADIUS' (1 (0x1))
        AVP: 'User-Name'(1) l=31 f=-M val="client1@eap.example.net"
        AVP: 'NAS-IP-Address'(4) l=12 f=-M val=<C0 A8 23 05>
        AVP: 'NAS-Identifier'(32) l=30 f=-M val="client.eap.testbed.aaa
    "
        AVP: 'NAS-Port'(5) l=12 f=-M val=1 (0x1)
        AVP: 'Called-Station-Id'(30) l=39 f=-M val="02-00-00-00-00-00:
    mac802
        AVP: 'Calling-Station-Id'(31) l=25 f=-M val="02-00-00-00-01-00"
        AVP: 'NAS-Port-Type'(61) l=12 f=-M val='Wireless - IEEE 802.11
    [RFC2
        AVP: 'Connect-Info'(77) l=30 f=-M val="CONNECT 54Mbps 802.11g"
        AVP: 'Class'(25) l=21 f=-M val=<66 44 2F 72 67 77 78 2F 61 61
    69 3A
        AVP: 'Acct-Session-Time'(46) l=12 f=-M val=279 (0x117)
        AVP: 'Accounting-Input-Packets'(365) l=16 f=-M val=54 (0x36)
        AVP: 'Accounting-Output-Packets'(366) l=16 f=-M val=17 (0x11)
        AVP: 'Accounting-Input-Octets'(363) l=16 f=-M val=10585 (0x2959
    )
        AVP: 'Accounting-Output-Octets'(364) l=16 f=-M val=2866 (0xb32)
        AVP: 'Event-Timestamp'(55) l=12 f=-M val=20160214T091043+03
SND to 'gw.eap.example.net':
    'Accounting-Answer'
      Version: 0x01
      Length: 196
      Flags: 0x40 (-P--)
      Command Code: 271
      ApplicationId: 3
      Hop-by-Hop Identifier: 0x148C55AE
      End-to-End Identifier: 0xB80F1463
        AVP: 'Session-Id'(263) l=82 f=-M val="client.example.net
    ;1455451008;4;client1@eap.example.net;gw.eap.example.net"
        AVP: 'Origin-Host'(264) l=30 f=-M val="server.eap.example.net"
        AVP: 'Origin-Realm'(296) l=23 f=-M val="eap.example.net"
        AVP: 'Result-Code'(268) l=12 f=-M val='DIAMETER_SUCCESS' (2001
    (0x7d
        AVP: 'Accounting-Record-Type'(480) l=12 f=-M val='STOP_RECORD'
    (4 (0
        AVP: 'Accounting-Record-Number'(485) l=12 f=-M val=3087996003
    (0xb80
```

Using such a gateway component it becomes possible to migrate a deployment as follows. We assume an existing infrastructure with many access points and one RADIUS server. First of all, create a new Diameter backend server and configure it to be able to authenticate the users from your realm. Create a second Diameter backend server if you are using accounting and configure it as well. Then set a RADIUS-Diameter gateway similar to the example above. Reconfigure your legacy access points to connect to the gateway, while your new access points may support Diameter directly and can be configured to connect to the backend directly. Finally, you may retire your RADIUS server and benefit from the flexibility and dynamic security of Diameter.

6.6 S6a Interface

The setup, maintenance, and release of the connection between the mobile phone, known as the UE in 3GPP parlance, and the mobile network is called mobility management. The Diameter application that handles mobility management is called the S6a interface and is part of the Evolved Packet Core (EPC), the all-packet network specified by 3GPP and introduced in Release 8.

6.6.1 Evolved Packet Core

The four main network elements of the EPC are the Mobility Management Entity, the Home Subscriber Server, the Serving Gateway, and the Packet Data Network Gateway. Figure 6.14 illustrates the EPC network architecture and its main Diameter-based signaling interfaces.

The Mobility Management Entity (MME) is the network element that provides UE authentication, network security, UE roaming capability, and selection of the UE's Serving Gateway and Packet Data Network Gateway. Generically, it is also referred to as a serving node in 3GPP specifications. The MME is a Diameter client node in the S6a application.

The MME accesses the Home Subscriber Server (HSS) for identity, authentication, and policy information about the UE's subscriber. The HSS is a Diameter server node in the S6a application.

The Serving Gateway (Serving GW) is the anchor point when the UE moves from one cell to another and when the UE moves from one 3GPP technology to another (2G to LTE, for example). The SGW routes user data packets and keeps track of UE session data.

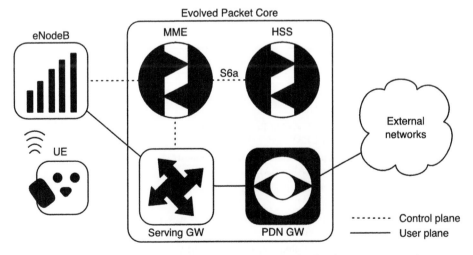

Figure 6.14 Evolved Packet Core. Connections to charging and policy functions are not shown.

The Packet Data Network Gateway (PDN GW) provides the UE with an entrance to and exit from the external packet data networks like the Internet. The PDN GW provides the UE an IP address and enforces operator policy, filters packets, and provides charging support, lawful interception, and packet screening.

6.6.2 S6a Overview

The S6a interface is the Diameter interface between the MME and the HSS. The S6a interface enables the transfer of subscriber-related (subscription and authentication) data between the HSS and the MME over SCTP. It is also used to communicate location changes (roaming) of the UE. Multiple Diameter commands are defined for this interface and are listed in Table 6.5. This application does not use the base Diameter accounting commands. This book only provides an overview of the S6a interface. For more information on the S6a Diameter commands and AVPs, see TS 29.272 [16]. Details of HSS and MME behavior can be found in TS 23.401 [17].

Note there is a very similar interface, known as S6d, between the HSS and the Serving GPRS Support Node (SGSN), which is responsible for the delivery of data packets to and from the mobile stations using older radio access technologies. The S6d interface uses the same Diameter commands as the S6a interface, but includes different AVPs.

The HSS and MME advertise support of the Diameter S6a application in CER/CEA commands by including the value of the S6a Application-Id (16777251) in the `Auth-Application-Id` AVP within the `Vendor-Specific-Application-Id` grouped AVP.

The 3GPP's vendor identifier (10415) is used in the `Supported-Vendor-Id` AVP and `Vendor-Id` AVPs within the `Vendor-Specific-Application-Id` grouped AVP. The manufacturer of a S6a Diameter node includes its ID in the top-level `Vendor-Id` AVP in the CER/CEA only, and not in the `Vendor-Specific-Application-Id` AVP.

Table 6.5 S6a interface application.

Reference	TS 29.272 [16], TS 23.401 [17]	
Application-Id value	16777251	
Commands	Update-Location-Request/Answer (ULR/ULA)	316
	Cancel-Location-Request/Answer (CLR/CLA)	317
	Authentication-Information-Request/ Answer (AIR/AIA)	318
	Insert-Subscriber-Data-Request/Answer (IDR/IDA)	319
	Delete-Subscriber-Data-Request/Answer (DSR/DSA)	320
	Purge-UE-Request/Answer (PUR/PUA)	321
	Reset-Request/Answer (RSR/RSA)	322
	Notify-Request/Answer (NOR/NOA)	323

6.6.2.1 Common AVPs for S6a Commands

The following AVPs are common to most S6a commands. Diving into the details of all S6a AVPs is beyond the scope of this book. For more information regarding S6a AVPs, consult TS 29.272 [16].

`Auth-Session-State`: Set to `NO_STATE_MAINTAINED` to indicate that the S6a Diameter sessions are implicitly terminated, which means the server does not maintain state information, and the client does not send re-authorization or session termination requests to the server. Neither the MME nor the HSS include the `Authorization-Lifetime` AVP nor the `Session-Timeout` AVP in their requests or responses.

`Experimental-Result`: While the `Result-Code` AVP is used to convey Diameter base protocol success or error messages, this AVP is used for S6a-specific errors. This grouped AVP contains the 3GPP Vendor ID in the `Vendor-Id` AVP, and the error code in the `Experimental-Result-Code` AVP. The error code conveyed can indicate transient failure at the HSS, unknown user, lack of subscription, and issues with the user equipment.

`Error-Diagnostic`: Provides more information about the error given in the `Experimental-Result` AVP. Diagnostics include details on the GPRS data and operator determined barring.

`Supported-Features`: Indicates the MME's and HSS's support for various network features. This mechanism ensures interoperability across various 3GPP releases. The MME and HSS can exchange feature support information in any of the S6a commands, although the exchange typically takes place with Update-Location-Request/Answer and Insert-Subscriber-Data-Request/Answer messages. As of Release 14, there are more than 50 features that can be supported. For some features, like operator-determined barring or regional subscriptions, if the MME does not indicate support, then the HSS may bar the UE from roaming. For other features, if the MME does not support them, the HSS merely stores this information and doesn't send any more information regarding the feature to the MME:

- Tracing
- LoCation Services
- Mobile-Originated Point-to-Point SMS
- Supplementary Service Barring
- UE Reachability and Attachment
- Local Time Zone Retrieval
- SMS in MME
- Proximity-based Services
- P-CSCF Restoration
- Reset-ID.

`Subscription-Data`: Contains the subscription profile of the user and is used with the Update-Location-Answer and Insert-Subscriber-Data-Request commands.

`Destination-Host`: Although the CCF for some of the Diameter S6a requests indicate that the `Destination-Host` AVP is optional, this is for compatibility with older 3GPP releases. As of Release 12, the `Destination-Host` AVP is mandatory for all S6a commands.

6.6.3 Authentication

6.6.3.1 Authentication-Information-Request Command

When an UE attaches to the network, it needs to be authenticated. If the UE has attached previously, the MME will request authentication parameters from the MME that handled the previous attachment to the network.[4] If the UE is new to the network, or if the security parameters for the UE's Attach Request are invalid, the MME uses S6a authentication commands to receive vectors from the HSS that the MME will use to perform authentication and key agreement. The MME requests authentication information for the subscriber from the HSS with the Authentication-Information-Request (AIR) (Figure 6.15).

Authentication-Information-Request AVPs

`Requested-EUTRAN-Authentication-Info`: A grouped AVP that contains information related to authentication requests for the radio access network used in LTE networks (E-UTRAN). The AVPs carry the number of requested authentication vectors that the MME can receive and whether the MME needs the vectors immediately (`Immediate-Response-Preferred` AVP). If the authentication synchronization fails on the UE, the MME includes re-synchronization information in the included `Re-Synchronization-Info` AVP.

`Visited-PLMN-ID`: Identifies the visited public land mobile network (PLMN).

`AIR-Flags`: A bit mask indicating that the MME requests the HSS to send the UE's usage type, which will enable the selection of a specific dedicated core network (DCN).

```
< AIR > ::= < Diameter Header: 318, REQ, PXY, 16777251 >
              < Session-Id >
              [ DRMP ]
              [ Vendor-Specific-Application-Id ]
              { Auth-Session-State }
              { Origin-Host }
              { Origin-Realm }
              [ Destination-Host ]
              { Destination-Realm }
              { User-Name }
              [ OC-Supported-Features ]
            * [ Supported-Features ]
              [ Requested-EUTRAN-Authentication-Info ]
              { Visited-PLMN-Id }
              [ AIR-Flags ]
            * [ AVP ]
            * [ Proxy-Info ]
            * [ Route-Record ]
```

Figure 6.15 Authentication-Information-Request Command Code Format.

---

4 MMEs communicate with each other using a non-Diameter protocol.

6.6.3.2 Authentication-Information-Answer Command

When the HSS receives the MME's Authentication-Information-Request, it checks the `Origin-Realm` AVP to ensure that the MME is allowed to use the information in the serving network given in the `Visited-PLMN-Id` AVP.

If the `Re-Synchronization-Info` AVP is present, the HSS must successfully verify the AUTS parameter before the HSS requests that the Authentication Centre (AuC) generate the corresponding requested authentication vectors. Fewer vectors than requested may be generated due to load and/or the presence of the `Immediate-Response-Preferred` AVP. If the HSS cannot calculate authentication vectors due to problems with provided UE information, the HSS returns the error `DIAMETER_AUTHORIZATION_REJECTED`. If the HSS experiences internal database errors, it responds to the MME with a `DIAMETER_AUTHENTICATION_DATA_UNAVAILABLE` error message. If the MME receives an error, it may re-attempt the Authentication-Information-Request.

The HSS sends the requested authentication vectors in its Authentication-Information-Answer (AIA) (Figure 6.16).

Authentication-Information-Answer AVPs

`Authentication-Info`: A Grouped AVP that only contains the `E-UTRAN-Vector` AVP when transmitted on the S6a interface. The `E-UTRAN-Vector` AVP is type Grouped and contains:

`Item-Number`: Used to order vectors received within one request. The MME should use the vector with the lowest Item-Number first.

`RAND`: Contains a RANDom number, a challenge.

`XRES`: Contains the eXpected RESponse.

`AUTN`: Contains the AUThentication tokeN.

```
< AIA > ::= < Diameter Header: 318, PXY, 16777251 >
            < Session-Id >
            [ DRMP ]
            [ Vendor-Specific-Application-Id ]
            [ Result-Code ]
            [ Experimental-Result ]
            [ Error-Diagnostic ]
            { Auth-Session-State }
            { Origin-Host }
            { Origin-Realm }
            [ OC-Supported-Features ]
            [ OC-OLR ]
          * [ Load ]
          * [ Supported-Features ]
            [ Authentication-Info ]
            [ UE-Usage-Type ]
          * [ AVP ]
          * [ Failed-AVP ]
          * [ Proxy-Info ]
          * [ Route-Record ]
```

Figure 6.16 Authentication-Information-Answer Command Code Format.

KASME: Contains the K_{ASME}, a key derived from a cipher key, integrity key, and serving network's identity.

UE-Usage-Type: Used to specify the usage characteristics of the UE. As of Release 14, there are no standardized values, only operator-specific ones.

The MME sends the challenge, authentication token, and a key set identifier KSI_{ASME},[5] which identifies the K_{ASME}, to the UE. The UE computes the expected message authentication code and compares it to the message authentication code within the authentication token [18]. If the comparison is successful, the UE computes a response and sends it to the MME. The MME checks that the response equals the expected response received from the HSS. If it does, the authentication is successful. If the authentication is not successful, the MME may either send another identity request to the UE or reject the UE's authentication request. More details on authentication can be found in TS 33.401 [19].

6.6.4 Location Management

The S6a interface commands Update-Location-Request/Answer (ULR/ULA) (command code 316) allow the MME to inform the HSS that it is currently serving an UE, and allow the HSS to update the MME with user subscription data.

6.6.4.1 Update-Location-Request Command

The MME sends a Update-Location-Request command (ULR) (Figure 6.17) to the HSS when one of the following scenarios happens:

- the UE attaches to the network for the first time,
- the UE has changed MMEs since it last attached to the network,
- the MME does not have valid subscription data for the UE,
- for some network sharing scenarios (e.g., where multiple operators share MMEs in a core network) if the PLMN-ID supplied by the network is different from that of the Globally Unique Temporary Identity of the UE.

Update-Location-Request AVPs

User-Name: Contains the user's identity (IMSI).

Terminal-Information: Contains information about the UE itself.

RAT-Type: Identifies the Radio Access Type (RAT) the UE is using.

ULR-Flags: A bit mask that indicates whether the UE's network attachment is initial, type of node (i.e., MME or SGSN) that sent the ULR, whether the HSS should send subscriber data to the MME, and whether the UE can support UE-based location services.

UE-SRVCC-Capability: Indicates whether the UE supports Single Radio Voice Call Continuity, which provides an interim solution for handing VoLTE traffic to 2G/3G networks.

Visited-PLMN-ID: Identifies the visited PLMN.

---

5 The KSI_{ASME} lets the UE and the MME identify a K_{ASME} without invoking the authentication procedure when the UE re-attaches.

```
< ULR > ::= < Diameter Header: 316, REQ, PXY, 16777251 >
               < Session-Id >
               [ DRMP ]
               [ Vendor-Specific-Application-Id ]
               { Auth-Session-State }
               { Origin-Host }
               { Origin-Realm }
               [ Destination-Host ]
               { Destination-Realm }
               { User-Name }
               [ OC-Supported-Features ]
            *  [ Supported-Features ]
               [ Terminal-Information ]
               { RAT-Type }
               { ULR-Flags }
               [ UE-SRVCC-Capability ]
               { Visited-PLMN-Id }
               [ Homogeneous-Support-of-IMS-Voice-Over-PS-Sessions ]
               [ GMLC-Address ]
            *  [ Active-APN ]
               [ Equivalent-PLMN-List ]
               [ MME-Number-for-MT-SMS ]
               [ SMS-Register-Request ]
               [ SGs-MME-Identity ]
               [ Coupled-Node-Diameter-ID ]
               [ Adjacent-PLMNs ]
               [ Supported-Services ]
            *  [ AVP ]
            *  [ Proxy-Info ]
            *  [ Route-Record ]
```

Figure 6.17 Update-Location-Request Command Code Format.

Homogeneous-Support-of-IMS-Voice-Over-PS-Sessions: Indicates homogeneous support. In order to route incoming IMS voice calls to the correct domain, the Terminating Access Domain Selection (T-ADS) requires homogeneous support/non-support of IMS voice over PS session for all of the UE's tracking areas [17]. If the support is either non-homogeneous or unknown, the MME does not include this AVP. The MME can send the AVP later when it has more information.

GMLC-Address: Contains the IPv4 or IPv6 address of the Visited Gateway Mobile Location Centre (V-GMLC) associated with the MME.

Active-APN: Contains the list of active access point names stored by the MME, including the identity of the PDN GW assigned to each APN. The MME may include it when the MME needs to restore PDN GW data in the HSS due to a reset procedure.

Equivalent-PLMN-List: Contains the equivalent PLMN list for which the MME requests the corresponding Closed Subscriber Group (CSG) Subscription data. A CSG identifies subscribers of an operator who are permitted only restricted access to one or more of the PLMN's cells.

MME-Number-for-MT-SMS: Included when the MME supports SMS and contains the ISDN number of the MME to route SMS to the UE.

SMS-Register-Request: Informs the HSS if the MME needs to be registered for SMS, prefers not to be registered for SMS, or has no preference.

SGs-MME-Identity: Informs the HSS of the MME identity that it uses over the SG interface, which is the interface to the Mobile Switching Center (MSC) that supports circuit-switched fallback.

Coupled-Node-Diameter-ID: A DiameterIdentity that is included if the requesting node is a combined MME/SGSN.

Adjacent-PLMNs: Type Grouped, contains a list of PLMN IDs where an UE is likely to make a handover from the PLMN where the MME is located.

Supported-Services: Type Grouped, contains monitoring events supported by the MME, such as UE reachability, UE location, roaming status, etc.

Note that currently none of the features in the Supported-Features AVP of the ULR command requires HSS support to successfully process the ULR command. For this reason the MME does not need to set the M-bit of the Supported-Features AVP in the ULR command.

If the UE is making an emergency attachment to the network, but the UE was not successfully authenticated, the MME does not send an Update-Location-Request to the HSS. The MME also ignores any unsuccessful Update-Location-Answer from HSS and continues with the attach procedure.

6.6.4.2 Cancel-Location-Request Command

When the HSS receives an Update-Location-Request, it first sends a Cancel-Location-Request (CLR) (Figure 6.18) with a Cancellation-Type of MME_UPDATE_PROCEDURE to the previous MME (if any) and replaces the stored MME Identity with the received value in the Origin-Host AVP. The HSS deletes any stored last-known MME location information.

```
< CLR > ::= < Diameter Header: 317, REQ, PXY, 16777251 >
            < Session-Id >
            [ DRMP ]
            [ Vendor-Specific-Application-Id ]
            { Auth-Session-State }
            { Origin-Host }
            { Origin-Realm }
            { Destination-Host }
            { Destination-Realm }
            { User-Name }
          * [ Supported-Features ]
            { Cancellation-Type }
            [ CLR-Flags ]
          * [ AVP ]
          * [ Proxy-Info ]
          * [ Route-Record ]
```

Figure 6.18 Cancel-Location-Request Command Code Format.

Cancel-Location-Request AVPs

Cancellation-Type: The following values indicate the reason for cancellation:

  0 MME_UPDATE_PROCEDURE: The HSS has received a ULR from a new MME, and is cancelling location with the MME receiving the CLR.

  1 SGSN_UPDATE_PROCEDURE: S6d interface only.

  2 SUBSCRIPTION_WITHDRAWAL: The HSS's operator has withdrawn the user's subscription.

  3 UPDATE_PROCEDURE_IWF: Used by an InterWorking Function when interworking with a pre-Release 8 HSS.

  4 INITIAL_ATTACH_PROCEDURE: The HSS has received an ULR for an Initial Attach procedure.

CLR-Flags: Indicates whether the MME needs to request that the UE re-attach.

6.6.4.3 Cancel-Location-Answer Command

The MME responds to the HSS's Cancel-Location-Request with a Cancel-Location-Answer (CLA) (Figure 6.19).

The Result-Code AVP contains Success and Permanent Failure result codes that are defined in the base protocol. The Experimental-Result AVP conveys S6a-specific failures.

6.6.4.4 Update-Location-Answer Command

After the HSS sends the Cancel-Location-Request to the old MME, the HSS acknowledges the Update-Location-Request from the new MME by sending the new MME an Update-Location-Answer (ULA) message (Figure 6.20).

If the Update-Location-Request includes information on one or more active APNs, the HSS deletes its stored PDN GW information, if any, and replaces it with the PDN GW information received in the list of Active-APN AVPs.

If the MME does not indicate support for an operator-determined barring feature or regional subscriptions in the Supported-Features AVP, the HSS will reject the MME's Update-Location-Request with a DIAMETER_ERROR_ROAMING_NOT_ALLOWED error.

```
< CLA > ::= < Diameter Header: 317, PXY, 16777251 >
              < Session-Id >
              [ DRMP ]
              [ Vendor-Specific-Application-Id ]
          *   [ Supported-Features ]
              [ Result-Code ]
              [ Experimental-Result ]
              { Auth-Session-State }
              { Origin-Host }
              { Origin-Realm }
          *   [ AVP ]
          *   [ Failed-AVP ]
          *   [ Proxy-Info ]
          *   [ Route-Record ]
```

Figure 6.19 Cancel-Location-Answer Command Code Format.

```
< ULA > ::= < Diameter Header: 316, PXY, 16777251 >
              < Session-Id >
              [ DRMP ]
              [ Vendor-Specific-Application-Id ]
              [ Result-Code ]
              [ Experimental-Result ]
              [ Error-Diagnostic ]
              { Auth-Session-State }
              { Origin-Host }
              { Origin-Realm }
              [ OC-Supported-Features ]
              [ OC-OLR ]
              [ Load ]
          *   [ Supported-Features ]
              [ ULA-Flags ]
              [ Subscription-Data ]
          *   [ Reset-ID ]
          *   [ AVP ]
          *   [ Failed-AVP ]
          *   [ Proxy-Info ]
          *   [ Route-Record ]
```

Figure 6.20 Update-Location-Answer Command Code Format.

Update-Location-Answer AVPs

ULA-Flags: A bit mask that is added when the Result-Code AVP is DIAME-
TER_SUCCESS and indicates how the HSS stores the MME information in memory
and whether the HSS has registered the MME for SMS.

When the MME receives a Update-Location-Answer, it validates the UE's presence
in the (new) tracking area. The MME can reject the UE's Attach Request if there are
restrictions in the user's regional subscription or access, or if there are subscription
check failures, such as the APN provided by the UE is not allowed by subscription. If
all checks are successful then the new MME constructs a context (basically, sets up the
UE's data session) for the UE.

6.6.5 Subscriber Data Handling

6.6.5.1 Insert-Subscriber-Data-Request Command

HSS sends an Insert-Subscriber-Data-Request (ISDR) (Figure 6.21) to replace or add
user data stored in the MME. The HSS can use the Insert-Subscriber-Data-Request message to inform the MME about the following:

- Operator changes to the user data, i.e., if the user's subscription to services has
  changed, or if the operator barred the user from roaming.
- Operator activates subscriber tracing.
- HSS requests notification from the MME when the UE becomes reachable.
- HSS requires data from the MME to support the Terminating Access Domain Selec-
  tion functionality.
- HSS wants location information and/or state information from the MME.

```
< ISDR > ::= < Diameter Header: 319, REQ, PXY, 16777251 >
                < Session-Id >
                [ DRMP ]
                [ Vendor-Specific-Application-Id ]
                { Auth-Session-State }
                { Origin-Host }
                { Origin-Realm }
                { Destination-Host }
                { Destination-Realm }
                { User-Name }
              * [ Supported-Features ]
                { Subscription-Data }
                [ IDR-Flags ]
              * [ Reset-ID ]
              * [ AVP ]
              * [ Proxy-Info ]
              * [ Route-Record ]
```

Figure 6.21 Insert-Subscriber-Data-Request Command Code Format.

- HSS wants the local time zone of the location in the visited network where the UE is attached.
- There was an update to the Session Transfer Number for Single Radio Voice Call Continuity (SRVCC), which provides an interim solution for handing over VoLTE to 2G/3G networks.
- HSS provides the MME with the identity of a dynamically allocated PDN GW as a result of the first PDN connection establishment associated with an APN over non-3GPP access.
- HSS has deregistered the MME for SMS.
- HSS has executed the P-CSCF restoration procedures, which are described in TS 23.380 [20].

Insert-Subscriber-Data-Request AVPs

IDR-Flags: A bit mask used by the HSS to indicate the information it is seeking about the UE, such as its attachment state, radio access type, location, and timezone. The HSS can also tell the MME that the MME has been unregistered for SMS, or that the HSS has started P-CSCF restoration procedures.

Subscription-Data: Contains the part of the subscription profile that is to be added to or replaced by the subscription profile stored in the MME.

6.6.5.2 Insert-Subscriber-Data-Answer Command

In response to the Insert-Subscriber-Data-Request from the HSS, the MME updates its stored subscription data and sends back an Insert-Subscriber-Data-Answer (ISDA) (Figure 6.22). Various AVPs are included if the HSS requested the information and if the information is available to the MME.

If the HSS indicated an operator-determined barring or regional subscriptions features in the Supported-Features AVP in its Insert-Subscriber-Data-Request, but

```
< ISDA > ::= < Diameter Header: 319, PXY, 16777251 >
               < Session-Id >
               [ DRMP ]
               [ Vendor-Specific-Application-Id ]
           *   [ Supported-Features ]
               [ Result-Code ]
               [ Experimental-Result ]
               { Auth-Session-State }
               { Origin-Host }
               { Origin-Realm }
               [ IMS-Voice-Over-PS-Sessions-Supported ]
               [ Last-UE-Activity-Time ]
               [ RAT-Type ]
               [ EPS-User-State ]
               [ EPS-Location-Information ]
               [ Local-Time-Zone ]
               [ Supported-Services ]
           *   [ Monitoring-Event-Report ]
           *   [ Monitoring-Event-Config-Status ]
           *   [ AVP ]
           *   [ Failed-AVP ]
           *   [ Proxy-Info ]
           *   [ Route-Record ]
```

Figure 6.22 Insert-Subscriber-Data-Answer Command Code Format.

the MME did not indicate support for it in its answer, the HSS may bar roaming and send a Cancel-Location-Request.

Insert-Subscriber-Data-Answer AVPs

IMS-Voice-Over-PS-Sessions-Supported: Indicates whether IMS Voice over PS sessions are supported by the UE's most recently used tracking area or routing area (0 NOT_SUPPORTED, 1 SUPPORTED). If the UE is in detached state, the MME does not include this AVP.

Last-UE-Activity-Time: Contains the time of the last radio contact with the UE. If the UE is detached from the network, this AVP is not included.

RAT-Type: Indicates the type of radio access technology the UE used during the last radio contact. If the UE is in detached state, this AVP is not included. Defined values are found in TS 29.212 [21].

EPS-User-State: Contains information about whether the UE is attached to the network.

EPS-Location-Information: Contains the UE's location.

Local-Time-Zone: Contains the local time zone and the daylight saving time adjustment of the location in the visited network where the UE is attached.

Supported-Services: Type Grouped, contains monitoring events supported by the MME, such as UE reachability, UE location, roaming status, etc.

Monitoring-Event-Report: Type Grouped, contains the report data for the monitoring event.

`Monitoring-Event-Config-Status`: Type Grouped, contains error details for the monitoring event.

6.6.5.3 Delete-Subscriber-Data-Request Command

The HSS uses the Delete-Subscriber-Data-Request (DSR) command (Figure 6.23) to remove the following data from the user profile stored in the MME:

- some or all of the EPS subscription data (Access Point Name Configuration Profile) for the subscriber
- the regional subscription
- the subscribed charging characteristics
- session Transfer Number for SRVCC
- trace data, which will cause the MME to deactivate the trace session
- ProSe subscription data
- Reset-IDs.

Delete-Subscriber-Data-Request AVPs

`DSR-Flags`: A bit mask that indicates what should be deleted from the subscriber data. Data to be deleted can include regional subscriptions, APN configurations, session information, and service information.

`SCEF-ID`: Type DiameterIdentity, contains the identity of the Service Capability Exposure Function (SCEF) that originated the service request towards the HSS.

`Context-Identifier`: Identifies the session information to be deleted and is included only if the `PDN subscription contexts Withdrawal` bit or the `PDP context withdrawal` bit is set in the `DSR-Flags` AVP.

```
< DSR > ::= < Diameter Header: 320, REQ, PXY, 16777251 >
              < Session-Id >
              [ DRMP ]
              [ Vendor-Specific-Application-Id ]
              { Auth-Session-State }
              { Origin-Host }
              { Origin-Realm }
              { Destination-Host }
              { Destination-Realm }
              { User-Name }
          *   [ Supported-Features ]
              { DSR-Flags }
              [ SCEF-ID ]
          *   [ Context-Identifier ]
              [ Trace-Reference ]
          *   [ TS-Code ]
          *   [ SS-Code ]
          *   [ AVP ]
          *   [ Proxy-Info ]
          *   [ Route-Record ]
```

Figure 6.23 Delete-Subscriber-Data-Request Command Code Format.

Trace-Reference: Contains the same value as used for the activation of the trace session. This element is present only if the Trace Data Withdrawal bit is set in the DSR-Flags AVP.

TS-Code: Contains the SMS-related teleservice codes to be deleted from the subscription. Included only if the SMS Withdrawal bit is set in the DSR-Flags AVP.

SS-Code: Contains the supplementary service codes to be deleted. Included only if the SMS Withdrawal bit or LCS Withdrawal bit is set in the DSR-Flags AVP.

6.6.5.4 Delete-Subscriber-Data-Answer Message

The MME responds to the HSS with a Delete-Subscriber-Data-Answer (DSA) (Figure 6.24), letting it know whether the data was successfully deleted.

6.6.6 Fault Recovery

The S6a interface commands Reset-Request/Answer (RSR/RSA) (command code 322) are used by the HSS to indicate to the MME that a failure has occurred. The RSR can also be used by the HSS as part of operation and maintenance actions to allow planned HSS outages without service interruption.

6.6.6.1 Reset-Request Command

The HSS sends Reset-Request (RSR) (Figure 6.25) to inform all relevant MMEs that the HSS has restarted and may have lost the current MME-Identity of some of its subscribers who may be roaming, and that the HSS, therefore, cannot send Cancel-Location-Requests or Insert-Subscriber-Data-Requests when needed.

Reset-Request AVPs

User-Id: Contains a list of User-Ids where a User-Id comprises the leading digits of an IMSI (i.e., MCC, MNC, leading digits of MSIN) and identifies the set of subscribers whose IMSIs begin with the User-Id. The HSS may include this AVP if the failure

```
< DSA > ::= < Diameter Header: 320, PXY, 16777251 >
            < Session-Id >
            [ DRMP ]
            [ Vendor-Specific-Application-Id ]
          * [ Supported-Features ]
            [ Result-Code ]
            [ Experimental-Result ]
            { Auth-Session-State }
            { Origin-Host }
            { Origin-Realm }
          * [ AVP ]
          * [ Failed-AVP ]
          * [ Proxy-Info ]
          * [ Route-Record ]
```

Figure 6.24 Delete-Subscriber-Data-Answer Command Code Format.

```
< RSR > ::= < Diameter Header: 322, REQ, PXY, 16777251 >
             < Session-Id >
             [ DRMP ]
             [ Vendor-Specific-Application-Id ]
             { Auth-Session-State }
             { Origin-Host }
             { Origin-Realm }
             { Destination-Host }
             { Destination-Realm }
           * [ Supported-Features ]
           * [ User-Id ]
           * [ Reset-ID ]
             [ Subscription-Data ]
             [ Subscription-Data-Deletion ]
           * [ AVP ]
           * [ Proxy-Info ]
           * [ Route-Record ]
```

Figure 6.25 Reset-Request Command Code Format.

is limited to subscribers identified by one or more User-Ids. Basically, a set of subscribers with identical leading IMSI digits.

Reset-ID: The HSS can include one or more Reset-ID AVPs identifying failed hardware components if the failure is limited to those subscribers associated with those components.

Subscription-Data: If the Reset-Request is to add or modify subscription data shared by multiple subscribers, this AVP contains the part of the subscription profile that is to be added to or to replaces the subscription profiles of the impacted subscribers stored in the MME.

Subscription-Data-Deletion: If the Reset-Request is to delete subscription data shared by multiple subscribers, this AVP identifies the part of the subscription profile that is to be deleted from the subscription profiles of the impacted subscribers stored in the MME.

6.6.6.2 Reset-Answer Command

When the MME receives a Reset-Request, it marks all impacted subscriber records. If the MME supports the Reset-ID AVP, it uses them together with the HSS's realm to determine which subscriber records are impacted. Otherwise the MME compares the HSS Identity received in the Origin-Host AVP with the value stored after successful ULA, and may use the received User-Id-List (if any) to determine which subscriber records are impacted.

At the next authenticated radio contact with the impacted UE, the restoration procedure is triggered.

The MME indicates success or failure to the HSS with a Reset-Answer (RSA) (Figure 6.26).

```
< RSA > ::= < Diameter Header: 322, PXY, 16777251 >
              < Session-Id >
              [ DRMP ]
              [ Vendor-Specific-Application-Id ]
            * [ Supported-Features ]
              [ Result-Code ]
              [ Experimental-Result ]
              { Auth-Session-State }
              { Origin-Host }
              { Origin-Realm }
            * [ AVP ]
            * [ Failed-AVP ]
            * [ Proxy-Info ]
            * [ Route-Record ]
```

Figure 6.26 Reset-Answer Command Code Format.

6.6.7 Notifications

The S6a interface commands Notify-Request/Answer (NOR/NOA) (command code 323) are used between the MME and the HSS when an inter-MME location update does not occur, but the HSS needs to be notified about the following updates:

- terminal information
- changes in UE SRVCC capability (if the MME supports SRVCC)
- an assignment/change of a dynamically allocated PDN GW for an APN, depending on access restrictions
- the HSS needs to send a Cancel-Location-Request to the current SGSN
- removal of MME registration for SMS
- when the SMS in MME feature is applied
- when the HSS has requested to be notified about when the UE is reachable, or if there is an update of the Homogeneous Support of IMS Voice Over PS Sessions.

6.6.7.1 Notify-Request Command

The MME sends a Notify-Request (NOR) (Figure 6.27) to notify the HSS about the following:

- an assignment or change of a dynamically allocated PDN GW for an access point name
- the need to send a Cancel Location when an inter-MME location update does not occur
- when the UE is reachable
- if the SMS in MME feature is applied, when the UE is reachable or the UE has memory capacity available to receive one or more short messages
- an update of the Homogeneous Support of IMS Voice Over PS Sessions
- the removal of MME registration for SMS.

Notify-Request AVPs

Terminal-Information: Notifies the HSS about changes to UE terminal information. Within this grouped AVP, only the IMEI and the Software-Version AVPs are used with the S6a interface.

```
< NOR > ::= < Diameter Header: 323, REQ, PXY, 16777251 >
              < Session-Id >
              [ Vendor-Specific-Application-Id ]
              [ DRMP ]
              { Auth-Session-State }
              { Origin-Host }
              { Origin-Realm }
              [ Destination-Host ]
              { Destination-Realm }
              { User-Name }
              [ OC-Supported-Features ]
          *   [ Supported-Features ]
              [ Terminal-Information ]
              [ MIP6-Agent-Info ]
              [ Visited-Network-Identifier ]
              [ Context-Identifier ]
              [ Service-Selection ]
              [ Alert-Reason ]
              [ UE-SRVCC-Capability ]
              [ NOR-Flags ]
              [ Homogeneous-Support-of-IMS-Voice-Over-PS-Sessions ]
              [ Maximum-UE-Availability-Time ]
          *   [ Monitoring-Event-Config-Status ]
              [ Emergency-Services ]
          *   [ AVP ]
          *   [ Proxy-Info ]
          *   [ Route-Record ]
```

Figure 6.27 Notify-Request Command Code Format.

MIP6-Agent-Info: This grouped AVP is included if a new PDN-GW has been selected, and the subscriber is allowed handover to non-3GPP access. It contains the identity of the PDN-GW as either an IP address or FQDN.

Visited-Network-Identifier: Contains the PLMN in which the PDN GW is located. The MME includes this when the PDN GW Identity is present but it does not contain an FQDN.

Context-Identifier: Identifies the APN configuration with which the selected PDN GW is correlated. The MME includes this AVP if it is available and the PDN-GW is present and is for a specific APN.

Service-Selection: Contains the APN for the selected and dynamically allocated PDN GW. This AVP is included if the selected PDN-GW is present and is for a specific APN.

Alert-Reason: Indicates if the subscriber is present, or if the handset has memory available to receive one or more short messages.

UE-SRVCC-Capability: Indicates whether the UE supports SRVCC capability.

NOR-Flags: A bit mask that can indicate whether HSS needs to send a Cancel-Location-Request, the UE has become reachable, if the UE can receive short messages via MME, if information regarding Homogeneous Support of IMS Voice Over PS Sessions has changed, or that the MME requests removal of its registration for SMS.

Homogeneous-Support-of-IMS-Voice-Over-PS-Sessions: This AVP is included if Homogeneous Support of IMS Voice Over PS Sessions has changed. If this AVP is not present, it indicates that there is no homogeneous support of this feature, or that the homogeneity of this support is unknown to the MME.

Maximum-UE-Availability-Time: Included to notify the HSS that the UE is reachable and provides a timestamp until which time a UE using a power-saving mechanism is expected to be reachable for Short Message delivery.

Monitoring-Event-Config-Status: Contains the status of those monitoring events whose configuration status has changed since the last status update.

Emergency-Services: Notifies the HSS that a new PDN-GW has been selected for the establishment of an emergency PDN connection.

6.6.7.2 Notify-Answer Command

The HSS responds to the MME with a Notify-Answer (NOA) (Figure 6.28) to indicate success or failure of the notification.

6.6.8 Ending Subscriber Sessions

When the user detaches from the network by turning off the phone or by being inactive for several days, the MME deletes the UE's subscription data and context and then informs the HSS via a Purge-UE-Request command (PUR) (Figure 6.29). Depending on implementation, the MME may delete UE data immediately after its detachment, or the MME may wait before purging, so that it can reuse the UE data when the UE attaches again without accessing the HSS.

6.6.8.1 Purge-UE-Request AVPs

PUR-Flags: A bit mask to indicate from which part of a combined MME/SGSN the UE was purged.

```
< NOA > ::= < Diameter Header: 323, PXY, 16777251 >
              < Session-Id >
              [ DRMP ]
              [ Vendor-Specific-Application-Id ]
              [ Result-Code ]
              [ Experimental-Result ]
              { Auth-Session-State }
              { Origin-Host }
              { Origin-Realm }
              [ OC-Supported-Features ]
              [ OC-OLR ]
           *  [ Load ]
           *  [ Supported-Features ]
           *  [ AVP ]
           *  [ Failed-AVP ]
           *  [ Proxy-Info ]
           *  [ Route-Record ]
```

Figure 6.28 Notify-Answer Command Code Format.

```
< PUR > ::= < Diameter Header: 321, REQ, PXY, 16777251 >
              < Session-Id >
              [ DRMP ]
              [ Vendor-Specific-Application-Id ]
              { Auth-Session-State }
              { Origin-Host }
              { Origin-Realm }
              [ Destination-Host ]
              { Destination-Realm }
              { User-Name }
              [ OC-Supported-Features ]
              [ PUR-Flags ]
          *   [ Supported-Features ]
              [ EPS-Location-Information ]
          *   [ AVP ]
          *   [ Proxy-Info ]
          *   [ Route-Record ]
```

Figure 6.29 Purge-UE-Request Command Code Format.

`EPS-Location-Information`: Contains the last known location information of the purged UE.

6.6.8.2 Purge-UE-Answer Command

When the HSS receives a Purge-UE-Request, it checks the user's IMSI. If the IMSI is known, the HSS sets the `Result-Code` to `DIAMETER_SUCCESS` and stores the last known location information given in the Purge-UE-Request message. It sends back to the MME a Purge-UE-Answer (PUA) command (Figure 6.30).

```
< PUA > ::= < Diameter Header: 321, PXY, 16777251 >
              < Session-Id >
              [ DRMP ]
              [ Vendor-Specific-Application-Id ]
          *   [ Supported-Features ]
              [ Result-Code ]
              [ Experimental-Result ]
              { Auth-Session-State }
              { Origin-Host }
              { Origin-Realm }
              [ OC-Supported-Features ]
              [ OC-OLR ]
          *   [ Load ]
              [ PUA-Flags ]
          *   [ AVP ]
          *   [ Failed-AVP ]
          *   [ Proxy-Info ]
          *   [ Route-Record ]
```

Figure 6.30 Purge-UE-Answer Command Code Format.

Purge-UE-Answer AVPs

`PUA-Flags`: Indicates that the MME should not immediately reuse the Temporary Mobile Subscriber Identity.

When the MME receives a successful Purge-UE-Answer from the HSS, it then checks the `PUA-Flags` AVP for the `Freeze M-TMSI` flag, and if set, it blocks M-TMSI for immediate reuse.

6.6.9 Extensibility

The S6a interface supports many features and is extensible. To preserve backwards-compatibility, any new functionality should be defined as optional. When extending the S6a application by adding new AVPs, the new AVPs should have the M-bit cleared and are defined as optional in the command code format.

However, if changes are unavoidably backwards-incompatible, the new functionality should be introduced as a new feature, and support for it advertised with the `Supported-Features` AVP. If the receiver must support the feature in order to process the request, then the feature is included in the `Supported-Features` AVP and the M-bit of the `Supported-Features` AVP is set.

References

1 V. Fajardo, J. Arkko, J. Loughney, and G. Zorn. Diameter Base Protocol. RFC 6733, Internet Engineering Task Force, Oct. 2012.

2 G. Zorn. Diameter Network Access Server Application. RFC 7155, Internet Engineering Task Force, Apr. 2014.

3 P. Calhoun, T. Johansson, C. Perkins, T. Hiller, and P. McCann. Diameter Mobile IPv4 Application. RFC 4004, Internet Engineering Task Force, Aug. 2005.

4 K. Sklower, B. Lloyd, G. McGregor, D. Carr, and T. Coradetti. The PPP Multilink Protocol (MP). RFC 1990, Internet Engineering Task Force, Aug. 1996.

5 3GPP. Telecommunication management; Charging management; Diameter charging applications. TS 32.299, 3rd Generation Partnership Project, Jan. 2018.

6 H. Hakala, L. Mattila, J.-P. Koskinen, M. Stura, and J. Loughney. Diameter Credit-Control Application. RFC 4006, Internet Engineering Task Force, Aug. 2005.

7 3GPP. Telecommunication management; Charging management; Charging architecture and principles. TS 32.240, 3rd Generation Partnership Project, Apr. 2011.

8 D. Sun, P. McCann, H. Tschofenig, T. Tsou, A. Doria, and G. Zorn. Diameter Quality-of-Service Application. RFC 5866, Internet Engineering Task Force, May 2010.

9 J. Korhonen, H. Tschofenig, M. Arumaithurai, M. Jones, and A. Lior. Traffic Classification and Quality of Service (QoS) Attributes for Diameter. RFC 5777, Internet Engineering Task Force, Feb. 2010.

10 J. Korhonen, H. Tschofenig, and E. Davies. Quality of Service Parameters for Usage with Diameter. RFC 5624, Internet Engineering Task Force, Aug. 2009.

11 L.-N. Hamer, B. Gage, and H. Shieh. Framework for Session Set-up with Media Authorization. RFC 3521, Internet Engineering Task Force, Apr. 2003.

12 W. Marshall. Private Session Initiation Protocol (SIP) Extensions for Media Authorization. RFC 3313, Internet Engineering Task Force, Jan. 2003.

13 L.-N. Hamer, B. Gage, B. Kosinski, and H. Shieh. Session Authorization Policy Element. RFC 3520, Internet Engineering Task Force, Apr. 2003.

14 P. Calhoun, G. Zorn, D. Spence, and D. Mitton. Diameter Network Access Server Application. RFC 4005, Internet Engineering Task Force, Aug. 2005.

15 P. Eronen, T. Hiller, and G. Zorn. Diameter Extensible Authentication Protocol (EAP) Application. RFC 4072, Internet Engineering Task Force, Aug. 2005.

16 3GPP. Evolved Packet System (EPS); Mobility Management Entity (MME) and Serving GPRS Support Node (SGSN) related interfaces based on Diameter protocol. TS 29.272, 3rd Generation Partnership Project, Jan. 2018.

17 3GPP. General Packet Radio Service (GPRS) enhancements for Evolved Universal Terrestrial Radio Access Network (E-UTRAN) access. TS 23.401, 3rd Generation Partnership Project, Apr. 2015.

18 3GPP. Digital cellular telecommunications system (Phase 2+); Universal Mobile Telecommunications System (UMTS); 3G security; Security architecture. TS 33.102, 3rd Generation Partnership Project, Jan. 2015.

19 3GPP. Digital cellular telecommunications system (Phase 2+); Universal Mobile Telecommunications System (UMTS); LTE; 3GPP System Architecture Evolution (SAE); Security architecture. TS 33.401, 3rd Generation Partnership Project, Oct. 2015.

20 3GPP. Digital cellular telecommunications system (Phase 2+); Universal Mobile Telecommunications System (UMTS); LTE; IMS Restoration Procedures. TS 23.380, 3rd Generation Partnership Project, May 2017.

21 3GPP. Universal Mobile Telecommunications System (UMTS); LTE; Policy and Charging Control (PCC); Reference points. TS 29.212, 3rd Generation Partnership Project, Apr. 2015.

7

Guidelines for Extending Diameter

7.1   Introduction

This chapter discusses an advanced topic of interest to those who have worked in standardization, have written Diameter implementations, or who have used Diameter to solve specific business problems. As discussed throughout this book, standardization efforts in Diameter have led to a number of specifications that cover a wide range of use cases. Still, you may have reached a point where the existing functionality does not cover your use case or it does not completely fulfill your requirements. You may have come to the conclusion that you need to extend the Diameter protocol. You are most likely asking yourself the following question: "Why cannot I extend the protocol in any way I want?"

To answer that question it is useful to provide some background information about the work in the Internet Engineering Task Force Diameter Maintenance and Extensions (DIME) working group. The working group suggests reusing as much as possible for the following reasons:

- The time to finalize a specification writing effort is significantly reduced since prior work can be taken into account. While it appears that writing specifications is fairly easy it still requires several months to years. The 80/20 rule applies: 80% of the time is spent on 20% of the specification.
- Reuse will lead to a smaller implementation effort. You will be able to make use of existing libraries (such as parsers). This does, however, require that your Diameter stack is written in modular fashion.
- The amount of testing (such as penetration testing) is significantly reduced since extensive testing has been conducted for existing modules and functionality already.

The most important reason, however, is to avoid interoperability problems, which was the main motivation for standardizing the Diameter protocol in the IETF. What does this mean? Imagine that you have Diameter deployed in your network (most likely involving products from different vendors) and, after some time, you would like to introduce another Diameter application into your network. Without following the extensibility guidelines outlined in the Diameter specification, problems may occur, such as the incorrect processing of messages. Imagine the following scenario where two companies independently develop a Diameter application and select the same Diameter Application-Id without registering these Application-Ids via the Internet Assigned Numbers Authority (IANA) registry. As a consequence, Diameter proxies and

Diameter: New Generation AAA Protocol – Design, Practice, and Applications, First Edition.
Hannes Tschofenig, Sébastien Decugis, Jean Mahoney and Jouni Korhonen.
© 2019 John Wiley & Sons Ltd. Published 2019 by John Wiley & Sons Ltd.

Diameter servers will be confused when they receive messages where they recognize the Application-Id but not the message's content. Even worse are cases where Diameter applications using the same Application-Id differ slightly in their functionality. With errors occurring only occasionally, these conditions will be difficult to diagnose.

Since you do not want these unnecessary problems, we give you the necessary background knowledge about Diameter extensibility.

7.2 Registration Policies

Before going into details of extending Diameter, we need to discuss the registration policies for extensions. At the end of a protocol design, you will most likely have new functionality that needs to be registered with IANA. The Diameter IANA registry contains pointers to AVPs, command codes, and Diameter applications. It not only collects a list of specifications developed in the IETF but also gathers specifications developed by other standards development organizations (SDOs). As such, it provides a wealth of information for supporting interoperability. The IANA registry webpage for Diameter can be found at http://www.iana.org/assignments/aaa-parameters/aaa-parameters .xhtml. Table 7.1 shows a snapshot of the Diameter Application-Id registry maintained by IANA.

In order to get your extension added to the IANA registry, certain requirements, often in terms of the amount of review, must be met. These requirements are called registration policies. The IETF has published a document listing a number of possible registration policies (RFC 5226 [1]), and for the purpose of this chapter the relevant policies are:

Standards Action: Values are only added to the registry with IETF Standards Track RFCs.

IETF Review: Values are only added to the registry through RFCs that have been shepherded through the Internet Engineering Steering Group (IESG) as Area Director-Sponsored or IETF working group documents.

Specification Required: Values and their meanings must be documented in a permanent and readily available public specification, in sufficient detail so that interoperability between independent implementations is possible. Specification Required also implies use of a Designated Expert, who will review the public specification and evaluate whether it is sufficiently clear to allow interoperable implementations.

Each section of the Diameter IANA registry has a specific registration policy associated with it. This policy dictates what requirements an extension has to meet in order for the requested code point to be added. This might give the impression that different code point numbers are "more valuable" than others since the difficulty for obtaining them varies. However, this segmentation of the available code space only aims to ensure that the space is not exhausted quickly while maintaining some flexibility. As long as a code point has been obtained for the respective Diameter extension, it does not matter from which range it came.

The amount of work needed to register a new Diameter extension varies with the policy. Standards Action, for example, typically requires more work than an IETF Review.

Table 7.1 Diameter Application-Id registry fragment.

ID value	Name	Reference
0	Diameter common message	[RFC6733]
1	NASREQ	[RFC7155]
2	Mobile IPv4	[RFC4004]
3	Diameter base accounting	[RFC6733]
4	Diameter Credit Control	[RFC4006]
5	Diameter EAP	[RFC4072]
6	Diameter SIP Application	[RFC4740]
...	...	...

More extensive review is useful when the available code space is close to exhaustion or if the extension has significant impact on other vendors. Note that registering your extension with IANA does not mean that you have to publish an RFC. You do not even have to publish an IETF specification at all. Various examples outlined in this book illustrate how standards development organizations and vendors have developed Diameter extensions and registered them with IANA. See, for example, Application-Id 16777251, which is assigned to the 3GPP S6a interface.

Sometimes, however, vendors do not want to make their specifications available nor are they willing to make any changes. In such a case, it is more useful to follow the path of vendor-specific extensions. Section 7.8 discusses vendor-specific extensions in detail and provides guidance for the interaction with IANA to make a smooth and quick registration.

7.3 Overview of Extension Strategies

To make it easier to add new features incrementally to the installed base, Section 1.3 of RFC 6733 [2] outlines a number of extension strategies. The Diameter design guidelines document, a companion document [3], covers the topic of extensibility in even greater detail.

Although reuse should be your guiding principle, we would like to point out that it should not be pushed too far. When the semantics of existing functionality differ from the envisioned new feature then it is better to define a new extension (even if it is largely a copy-and-paste of an already existing specification text).

The Diameter extensions are a direct consequence of the Diameter layers, as shown in Figure 7.1. The outermost layer in Diameter is the application concept, which encapsulates everything else. The next layer inside is the command. Finally, the innermost layer is the AVP. We start our description with the least intrusive and simplest extension, the AVP. Next, the extensions of commands are explained followed by applications.

An important design decision for extending Diameter is whether Diameter nodes are required to understand the new functionality or whether the newly defined features are purely optional. Whenever it is mandatory for Diameter clients, Diameter agents (except

Figure 7.1 Diameter extensions.

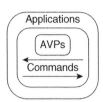

for relays and redirect agents) or Diameter servers to understand new functionality, a new Diameter application has to be defined.

7.4   Extending Attribute–Value Pairs

7.4.1   Extending Existing AVPs

Most specification authors define AVPs such that they can to be extended. Multiple ways for providing these extension points are possible, namely:

- allocating new AVP flags
- utilizing AVP extension points
- new values for use in AVPs.

7.4.1.1   Creating New AVP Flags

As described in Section 2.6.3 the header of an AVP defines an AVP flags field, which contains 8 bits. These AVP flags are not tied to a specific AVP but are applicable to all AVPs, even those defined in the past. Three flags have been registered already, namely the "V" bit (to denote vendor-specific AVPs), the "M" bit (for mandatory AVPs), and the "P" bit (for protected AVPs). Due to the limited AVP flags space in the header, the policy for creating new flags requires "Standards Action".

7.4.1.2   Adding AVP Extension Points

The data format for an AVP may either use a basic data format or a data format derived from other AVPs. The data format of type "Grouped" is special since it allows nesting of AVPs and, when appropriate extension points are defined, allows further sub-AVPs to be added to an existing AVP. As an example, consider the From-Spec AVP defined in the traffic classification and quality of service (QoS) attributes for Diameter (RFC 5777 [4]) shown in Figure 7.2.

The last line in Figure 7.2 contains the * [AVP] directive that offers the ability for future specifications to add any other AVP. Consequently, you can use this extension point to enhance the functionality of this AVP. When you design a new Grouped AVP, you should include this extension point.

7.4.1.3   Adding New AVP Values

AVPs may allow additional values to be added in future specifications. Typically this type of functionality has to be explicitly defined together with information about who is authorized to register new values and under what conditions.

```
From-Spec ::= < AVP Header: 515 >
             * [ IP-Address ]
             * [ IP-Address-Range ]
             * [ IP-Address-Mask ]
             * [ MAC-Address ]
             * [ MAC-Address-Mask]
             * [ EUI64-Address ]
             * [ EUI64-Address-Mask]
             * [ Port ]
             * [ Port-Range ]
               [ Negated ]
               [ Use-Assigned-Address ]
             * [ AVP ]
```

Figure 7.2 Traffic classification and QoS attributes for Diameter.

As an example, consider RFC 5777 again, which defines the `Treatment-Action` AVP. As explained in Section 10.3 of RFC 5777 a new registry is created and maintained by IANA that contains an initial list of four values: 0 for drop, 1 for shape, 2 for mark, and 3 for permit. New values can be added by anyone as long as they meet the Specification Required registration policy.

The policy for allocating new values for the most of the important AVPs in RFC 6733 has been changed to IETF Review. For example, a relevant AVP for protocol designers is the `Result-Code` AVP, which is used for error handling.

7.5 Extending Commands

Most Diameter commands are specified so that they can to be extended. Multiple ways for extending commands are possible, namely:

- allocating new Command flags
- adding AVPs
- creating new Commands.

7.5.1 Allocating New Command Flags

Section 3 of RFC 6733 describes the existing Command Flags field, and Table 2.3 summarizes the current set. Due to the limited space, the Command Flags can only be extended using Standards Action.

7.5.2 Adding New AVPs

When creating a new AVP the first design decision relates to the semantics of the AVP and its appropriate data type. The choice of data type for AVPs is important since the use of a structured data format allows for simpler parser implementations. As such, using a data format of OctetString to embed complex structures is likely to make it unnecessarily complex for an implementer to develop a parser. It may also lower

```
NAT-Internal-Address ::= < AVP Header: 599 >
                            [ Framed-IP-Address ]
                            [ Framed-IPv6-Prefix ]
                            [ Port]
                        *   [ AVP ]
```

Figure 7.3 `NAT-Internal-Address` AVP.

reusability and may make extensibility more difficult. As an example, consider the `IPFilterRule` AVP originally defined in the Diameter base protocol. The `IPFil-terRule` AVP uses OctetString as the basic data type. The content is, however, quite complex and contains a subset of **ipfw** functionality from FreeBSD. Compare the structure to RFC 5777 [4], which defines an extended version of the `IPFilterRule` AVP functionality using a Diameter-native format. The Diameter format allows for easier extensibility within Diameter, as demonstrated with [5], which aims to extend it for support of Explicit Congestion Notification (ECN).

When specifying Grouped AVPs, existing AVPs can be reused in your new Grouped AVP. The full pool of Diameter AVPs is available at your disposal. For example, RFC 6736 [6] defines a Grouped AVP called `NAT-Internal-Address` (Figure 7.3). The three optional AVPs that it contains are defined elsewhere. The `Framed-IP-Address` AVP and the `Framed-IPv6-Prefix` AVP are reused from the Diameter Network Access Server Application [7]. The `Port` AVP is reused from the traffic classification and QoS attributes for Diameter [4]. Consequently, this new AVP is created by reusing already available AVPs. Grouped AVPs may have a mix of settings of the "M" (mandatory) bit. The Grouped AVP may not have the "M" bit set, but an encapsulated AVP may have. The receiver of the Grouped AVP may simply ignore the encapsulated AVP if it does not recognize the Grouped AVP. The AVP Code numbering space of all AVPs included in a Grouped AVP is the same as for non-Grouped AVPs.

The AVP Code field is a 32-bit field, which is controlled and managed by IANA. The AVP Code 0 is not used since it represents the IETF AVP code space. AVP Codes 1–255 are managed separately as RADIUS Attribute Types, which allows RADIUS attributes to be reused by Diameter. For example, RFC 6736 [6] specifies the Diameter Network Address and Port Translation Control Application and it reuses an already standardized `Egress-VLANID` AVP from RFC 4675 [8]. The main purpose of RFC 4675 is to standardize RADIUS attributes, but it also registers the respective Diameter AVPs as part of the RADIUS/Diameter compatibility mechanism. Look at Section 4 of RFC 4675 [8] for an example of how RADIUS specifications in the past often registered Diameter AVPs. Note that registering Diameter AVPs through RADIUS specification is now discouraged.

AVPs may be allocated following Expert Review with Specification Required. A block allocation, which is the release of more than three AVPs at a time for a given purpose, requires IETF Review.

An interesting feature of the AVP structure is that it also allows for vendor-specific AVPs. Vendors can have their own AVP codes namespace that will be identified by their Vendor-ID (also known as the Enterprise-Number), and they control the assignments of their vendor-specific AVP codes within their own namespace. Since the 32-bits of the AVP code offer insufficient space for the Vendor-ID an additional field is included in the AVP header. This optional Vendor-ID field in the AVP header is present if the "V" bit is set in the AVP Flags field.

As such, the absence of a Vendor-ID field identifies the IETF AVP codes namespace and the presence of the Vendor-ID field indicates a vendor extension.

7.5.2.1 Adding New AVPs to Base Commands

It is possible to extend both CER and CEA commands with arbitrary optional AVPs. Based on the presence of specific AVPs in the CEA, the transport connection initiator knows whether the responder peer understood the extension(s). However, since there is no mechanism in the base protocol to determine whether a peer can understand the optional AVPs, there is no guarantee that the other peer understood them. If it is necessary to extend the capability exchange, then a lightweight, one round-trip negotiation protocol could be designed on top of the capability exchange.

RFC 3588 did not specify extensions for DPR/DPA or DWR/DWA. RCF 6733 "fixes" this by adding optional AVPs to the DPR/DPA and DWR/DWA CCF (* [AVP] extension). However, in order to avoid backward compatibility issues, extending DPR/DPA and DWR/DWA is not recommended. A compliant RFC 3588 implementation may discard a message or respond with an error if it determines that the CCF of the received request is invalid from the optional AVPs point of view.

7.5.3 Creating New Commands

A new command requires a unique command code.

The values 0–255 are reserved for RADIUS backward compatibility and are mapped to the RADIUS Packet Type Codes, which are also maintained by IANA at http://www.iana.org/assignments/radius-types/radius-types.xhtml.

The values 256–8,388,607 (0x100 to 0x7fffff) are for permanent, standard commands, allocated by IETF Review.

The values 8,388,608–16,777,213 (0x800000–0xfffffd) are reserved for vendor-specific command codes, to be allocated on a first come, first served basis by IANA.

The values 16,777,214 and 16,777,215 (hexadecimal values 0xfffffe–0xffffff) are reserved for experimental commands. As these codes are only for experimental and testing purposes, there is no guarantee of interoperability between Diameter peers using experimental commands.

7.5.3.1 Routing AVPs

The Route-Record AVP is of type DiameterIdentity and contains the identity of the Diameter node (i.e., the agent) that inserted the AVP. The Route-Record AVP is used only in request messages, and its content must be the same as the Diameter node's Origin-Host AVP that is used in the CER message during the capability exchange. A Diameter request message CCF should contain the following definition for the Route-Record AVP.

```
* [ Route-Record ]
```

A Diameter application may define its CCF so that it does not contain Route-Record AVPs or any routing AVPs. However, such a message must not be routed because a Diameter relay agent would add routing AVPs to any request message that it routes, independent of the application and its CCF. After traversing a relay, the request message CCF would be invalid, and any Diameter node performing CCF correctness verification would return a *permanent error* with the `Result-Code` AVP value set to the `DIAMETER_AVP_NOT_ALLOWED`.

7.6 Creating New Applications

What is the line between just extending an AVP or command and having to define a new Diameter application? When your Diameter extension does at least one of the following, then you have created a new application:

- Adds an AVP with the "M" bit in the MUST column or in the MAY column to an existing command or application.
- Adds a new command to an existing application.
- Modifies an existing command such that a new command code has to be registered. Note that removing the `Error-Message` AVP from a command's CCF does not count as a modification since the `Error-Message` AVP is for human consumption only and is not parsed by Diameter nodes.
- Deletes a command from an existing application.
- Modifies the meaning or semantics of an existing application's AVP Flags.
- Adds a new flag bit.

If your extension just adds optional AVPs with the "M" bit cleared to a command that has a CCF definition containing `*  [ AVP ]` then you do not need to define a new application.

7.6.1 The Application-Id

Your new application should have a unique Application-Id that is carried by all session-level messages, including the base commands (i.e., Re-Auth-Request/Answer, Session-Termination-Request/Answer, Abort-Session-Request/Answer). If your new application uses the base Accounting-Request/Answer commands, it should implement the coupled accounting service (Section 6.2.3) and use its own Application-Id in the accounting commands. Note that some existing specifications do not adhere to this rule for historical reasons. For example, the Rf interface application (Section 6.2.10) implemented mandatory to understand AVPs, and removed optional but mandatory to understand AVPs (e.g., the `Accounting-Realtime-Required` AVP). Because it did so, the Rf interface really should have had a new Application-Id rather that reusing the Application-Id of 3, which is the Application-Id of the base accounting application.

If you still want to use the split accounting service of the base protocol, the specification for extending Diameter [3] recommends that the `Auth-Application-ID` AVP, not the `Acct-Application-Id` AVP, be used as a way of identifying the application, even though it has also been traditionally a vendor-specific AVP.

Note that there is no versioning support provided by Application-Ids. Also, every Diameter application is a standalone application. The base protocol does not support linking Diameter applications.

Values from the range 0x00000001 to 0x00ffffff can be requested with the Specification Required policy. Values from the 0x01000000–0xfffffffe range are allocated for vendor-specific applications, and are obtained on a first-come, first-served basis.

7.7 Lessons Learned

From our own experience co-authoring Diameter specifications we would like to offer you ten lessons for your work on Diameter specifications.

1. UNDERSTAND THE BASIC DIAMETER CONCEPTS
   The most important starting point for your work on Diameter is to understand it first before writing your own specification. While this suggestion may seem obvious, we recognize that some people who write technical specifications do it as a learning-by-doing exercise before understanding the Diameter base specification. You will save yourself and others a lot of time by doing your homework first. If you have read the earlier sections of the book you are familiar with the concepts already and you should be doing fine.

2. YOU DON'T NEED TO USE THE CCF SYNTAX
   When you read the IETF Diameter specifications, you will notice that many of them define Diameter AVPs and Diameter commands using the Command Code Format (CCF) syntax. We provide an overview of the syntax in Section 2.6.1 and the formal definition of the syntax is given in the base specification [2]. It is, however, not a requirement to use CCF in your specification if you do not feel comfortable with it. In fact, the formal notation of CCF does not capture the nuances of protocol specification. For example, the specification text, and not the CCF, provides the information on when to use or not to use optional AVPs. The CCF syntax by itself is often insufficient to express the syntax of the commands. A good description can be better than correct CCF. If you feel uncomfortable writing CCF or if you readers are likely unfamiliar with it, you don't need to use it.

3. OFFER EXAMPLES
   Many readers of specifications will look at examples to better understand your specification. Offering example message flows and example AVP instantiations will help them. Most likely the examples will also help you to verify whether every feature has indeed been included in the specification. Make sure that your examples are practical and realistic, which will often require some information about the network setup. Also, make sure your examples are correct. Readers have been known to follow examples more closely than specification text.

4. DON'T FORGET THE ERROR CASES
   Your initial version of a specification is most likely very simple. Once additional use cases and requirements from stakeholders are included, the complexity will likely increase. In many cases all attention is focused on the positive use cases, which are the most likely ones. However, it is important to explore error situations as well to ensure that your protocol behaves properly when problems occur. Even if the error is not automatically recoverable, make sure that useful human-readable error messages are returned so that developers are able to diagnose the problem quickly. In many cases, covering error cases requires you to have implementation experience

with the specification you are developing. If you can create an implementation prototype, it will provide valuable feedback to the specification.

5. REUSE AS MUCH AS POSSIBLE

When you start writing your first specification it is often difficult to get started. You might be wondering how to translate your idea to an interoperable, technical specification. We recommend reusing an existing specification and letting it serve as a template. Of course, you will have to find a suitable starting point but Chapter 6 will walk you through a number of important specifications. It is common practice to reuse text from available RFCs.

6. REGISTER YOUR EXTENSION WITH IANA

Once you are finished with your Diameter protocol extension defining new AVPs, new Command Codes, or Diameter applications, register them with IANA. In the past various specification authors were reluctant to standardize Diameter applications since the allocation policy was rather strict. However, the registration policy was changed with the publication of RFC 5719 [9], and the policy was later incorporated into RFC 6733. Consequently, it became much easier to register AVPs, command codes, and Diameter applications. The benefit of registering extensions is that other vendors will know about the allocated code point and interoperability problems between two independent implementations that use the same code point will be avoided.

7. NOT EVERY FEATURE NEEDS TO BE REGISTERED WITH IANA

Despite the advice given in item 6, some vendors may be reluctant to register their Diameter protocol extension. Still, it makes a lot of sense to distinguish proprietary functionality from different vendors, particularly when telecommunication operators use Diameter in a multi-vendor environment. The Diameter base specification allows vendor-specific extensions, permitting vendors to allocate code points for their Diameter extensions without having to register them with IANA and without having to publish their specifications. Section 7.8 explains how vendor-specific extensions can be specified.

8. THINK ABOUT EXTENSIBILITY IN YOUR OWN APPLICATIONS

You may think that you are covering all the functionality that you need when designing your Diameter extension. However, if your protocol design will be used widely, then new uses of your extension will likely emerge, and your extension may need to be extended further. As such, you will have to consider adding extension points to your application in the same way as you may have used them yourself. At a minimum, you should you consider allowing AVPs to be extended (by using data types that allow new values to be added and by the use of grouped AVPs with extension points). Additionally, it is highly recommended to also offer extension points with command codes. Designing for future use is, however, an art that requires the specification author to have a vision about the future use of the specification that goes beyond the requirements list derived from its current use cases.

9. DUAL STRATEGY: DEFINING A DIAMETER APPLICATION AND AVPs

When developing a new extension, it can be difficult to decide whether a new application should be defined or whether an existing application should be extended via a set of optional AVPs. In fact, the book authors have encountered this situation several times, and have concluded that sometimes it is better to define both.

For example, RFC 5866 [10] defines a Diameter quality of service (QoS) application (Section 6.4), but RFC 5624 [11] provides QoS parameters that can be incorporated into other applications, like the Diameter Extensible Authentication Protocol (EAP) application [12]. The same approach has also been applied to Diameter Mobile IPv6, as described in RFC 5447 [13] and RFC 5778 [14]. RFC 5778 defines two Diameter applications, one for use with the custom Mobile IPv6 security protocol and the other for use with the Internet Key Exchange protocol. RFC 5447, on the other hand, only defines AVPs, which can be used with existing Diameter applications, such as Diameter EAP [12] and the Diameter Network Access Server Application [7]. While the design of the two specifications is quite different, various AVPs are shared by the two specifications.

10. TEST FOR EXTENSIBILITY

Specifying extensibility in a specification is clearly important. However, it is not enough. Interoperability testing is even more important to ensure that the implementation does indeed conform to the specification. Interoperability testing ensures that functionality is verified. Testing how the Diameter protocol stack reacts when confronted with unknown extensions is often forgotten. This has, in other protocols (such as TLS), caused problems in real-world deployments.

7.8   Vendor-specific Extensions

This section describes vendor-specific extensions and the steps to take to use them. Vendor-specific extension makes use of Enterprise Numbers (also known as Structure of Management Information (SMI) Network Management Private Enterprise Codes). These Enterprise Numbers are allocated by IANA and be found at http://www.iana.org/assignments/enterprise-numbers.

It is likely that your company is already on this list since this registry is used not only for Diameter but also for a number of other protocols (such as network management protocols). If your company does not yet have an Enterprise Number then one may be requested via the online form available at http://pen.iana.org/pen/PenApplication.page.

As described earlier in this chapter, not all Diameter extensions, for example AVP and Command flags, are suitable for vendor-specific extensions. The process for using vendor-specific AVPs, command codes, and Diameter applications is slightly different and described in the subsequent sub-sections.

7.8.1   AVPs

Vendors can have their own AVP codes namespace identified by their Vendor-ID, which corresponds to the Enterprise Number, and they control the assignments of their vendor-specific AVP codes within their own company. A vendor is then responsible for managing its AVP code space to avoid collisions. While this does not sound difficult, it becomes complicated when considering larger, multinational enterprises. These companies often have multiple teams working on independent Diameter products. When a vendor-specific AVP is implemented by more than one vendor, allocation of global AVPs should be encouraged instead.

7.8.2   Command Codes

The approach for registering vendor-specific Command Codes is different from the approach used for AVPs. Vendors have to request a Command Code from IANA in the respective range, as described in Section 7.5. Consequently, vendors have to send an email to IANA at iana-prot-param@iana.org to request the allocation of a Command Code value. The email needs to contain the name of the Command Code request/answer pair, optionally an abbreviation for the request/answer, and contact information about the person submitting the request (if not obvious from the email address itself).

7.8.3   Diameter Applications

The allocation of vendor-specific Diameter Application-Ids is similar to the procedure for registering command codes. Again, the code space has been separated into different ranges, and an email has to be sent to IANA at iana-prot-param@iana.org with the information about the name of the Diameter application together with information about the person submitting the request. IANA will allocate a number when processing the request.

As can be seen, registering vendor-specific extensions is easy and painless. Even the interaction with IANA is, from our own experience, fast and pleasant. For all vendors who have concerns about the need to disclose company-internal specifications it should be clear from the description in this section that this fear is unjustified.

7.9   Prototyping with `freeDiameter`

A core belief of the IETF is "rough consensus and running code". This is to highlight the importance of testing new ideas with real running code. `freeDiameter` has been created in particular to allow implementers to test their new ideas/applications relatively easily, without the need to worry about the base protocol implementation. The *test_app.fdx* extension can serve as a model of client-server implementation to build upon, or for those who are not familiar with C language, the *dbg_interactive.fdx* extension can be a convenient tool to getting started quickly. Please refer to Appendix B for detailed information on these extensions.

References

1 T. Narten and H. Alvestrand. Guidelines for Writing an IANA Considerations Section in RFCs. RFC 5226, Internet Engineering Task Force, May 2008.
2 V. Fajardo, J. Arkko, J. Loughney, and G. Zorn. Diameter Base Protocol. RFC 6733, Internet Engineering Task Force, Oct. 2012.
3 L. Morand, V. Fajardo, and H. Tschofenig. Diameter Applications Design Guidelines. RFC 7423, Internet Engineering Task Force, Nov. 2014.
4 J. Korhonen, H. Tschofenig, M. Arumaithurai, M. Jones, and A. Lior. Traffic Classification and Quality of Service (QoS) Attributes for Diameter. RFC 5777, Internet Engineering Task Force, Feb. 2010.

5 L. Bertz, S. Manning, and B. Hirschman. Diameter Congestion and Filter Attributes. RFC 7660, Internet Engineering Task Force, Oct. 2015.

6 F. Brockners, S. Bhandari, V. Singh, and V. Fajardo. Diameter Network Address and Port Translation Control Application. RFC 6736, Internet Engineering Task Force, Oct. 2012.

7 G. Zorn. Diameter Network Access Server Application. RFC 7155, Internet Engineering Task Force, Apr. 2014.

8 P. Congdon, M. Sanchez, and B. Aboba. RADIUS Attributes for Virtual LAN and Priority Support. RFC 4675, Internet Engineering Task Force, Sept. 2006.

9 D. Romascanu and H. Tschofenig. Updated IANA Considerations for Diameter Command Code Allocations. RFC 5719, Internet Engineering Task Force, Jan. 2010.

10 D. Sun, P. McCann, H. Tschofenig, T. Tsou, A. Doria, and G. Zorn. Diameter Quality-of-Service Application. RFC 5866, Internet Engineering Task Force, May 2010.

11 J. Korhonen, H. Tschofenig, and E. Davies. Quality of Service Parameters for Usage with Diameter. RFC 5624, Internet Engineering Task Force, Aug. 2009.

12 P. Eronen, T. Hiller, and G. Zorn. Diameter Extensible Authentication Protocol (EAP) Application. RFC 4072, Internet Engineering Task Force, Aug. 2005.

13 J. Korhonen, J. Bournelle, H. Tschofenig, C. Perkins, and K. Chowdhury. Diameter Mobile IPv6: Support for Network Access Server to Diameter Server Interaction. RFC 5447, Internet Engineering Task Force, Feb. 2009.

14 J. Korhonen, H. Tschofenig, J. Bournelle, G. Giaretta, and M. Nakhjiri. Diameter Mobile IPv6: Support for Home Agent to Diameter Server Interaction. RFC 5778, Internet Engineering Task Force, Feb. 2010.

Appendix A

`freeDiameter` Tutorial

A.1 Introduction to Virtual Machines

In the real world, Diameter applications are deployed on multiple machines within a single organization or across different organizations and networks. Because such a multi-machine environment is expensive to set up and challenging to maintain, this book's examples use an environment based on virtual machines.

A virtual machine (VM) is an emulation of a complete computer system with its own operating system and applications. A VM is created using virtualization software that runs on a physical machine. The virtualization software allows for quick and easy configuration of the VM, for example adding a new network interface is just a matter of a few clicks. The interface between the VM (called the *guest*) and the system running the virtualization software (your computer, the *host*) is clearly delimited, so there is no problem, for example, for a machine running Microsoft Windows 10 to host a guest that is running Linux and another guest that is running FreeBSD. Using such a VM environment, we can create the network between Diameter peers for our tests, without the need for several physical computers and network equipment. All the examples in the book can run easily on a regular office laptop. However, we recommend that you use a large screen, or several screens, in order to display the multiple VM screens simultaneously.

We have pre-configured a VM image containing `freeDiameter` to reduce your work effort, and we will walk you through the steps to run the examples. At a high level, these steps are as follows:

- Create your first VM from our pre-built image.
- Use this master VM as a template to create the other machines used in the different examples throughout the book via cloning, a simple operation performed through virtualization software.
- Issue a few commands on each VM to set their network configuration and to start `freeDiameter` with the correct profile.

Our VM runs under the Ubuntu operating system. We have chosen Ubuntu mostly because it seems more (new-)user-friendly. For more information about working with Ubuntu, please visit https://help.ubuntu.com.

Diameter: New Generation AAA Protocol – Design, Practice, and Applications, First Edition.
Hannes Tschofenig, Sébastien Decugis, Jean Mahoney and Jouni Korhonen.
© 2019 John Wiley & Sons Ltd. Published 2019 by John Wiley & Sons Ltd.

A.2   Installing the Virtualization Software

We have chosen *Oracle VirtualBox* as the virtualization software application. VirtualBox, which is freely available and can be used on Windows, OS X, Linux, and Solaris, should allow you to run all the examples without having to go through painful operating system installations and network configurations.

To install Oracle VirtualBox, visit https://www.virtualbox.org and follow the instructions to retrieve the correct package for your environment and install the application.

You may also use another virtualization software application that supports Open Virtualization Format (OVF) to launch the provided VM.

A.3   Creating Your Own Environment

If you do not want to or cannot use the provided VM environment, you can set up your own environment. Visit the official `freeDiameter` website at http://www.freediameter.net where you can find the source code and guidance for building and configuring `freeDiameter`. Note that while `freeDiameter` source code can be compiled easily on most POSIX systems, it relies on several libraries for specific operations, for example the GNU TLS library for the cryptography, the GNU IDN library for international domain names support, etc. As a result, `freeDiameter` can only be used (with reasonable effort) on systems that provide these dependencies within their package system. Fortunately, most modern operating systems provide the necessary packages.

The building and configuration information provided by the `freeDiameter` website can also be useful if you want to develop your own Diameter applications on top of this framework. Along with the examples you will also find a description of the network topology used in the respective tests. We believe that this will help you to recreate the setup in a different environment. This book is, however, not a tutorial for configuring IP-based networks. If you want to go beyond the examples described in the book you will have to consult books covering IP fundamentals.

A.4   Downloading the VM Image

Download the VM image `freeDiameter.ova` from this book's website at https://diameter-book.info. The file is rather large (2 GB) and may take some time to download.

A.5   Installing and Starting the Master VM `freeDiameter`

1. Start the *VirtualBox* application. The **Oracle VM VirtualBox Manager** screen appears.
2. From the **File** menu, select **Import Appliance**. The **Appliance to import** dialog box appears. Locate the `freeDiameter.ova` file that you downloaded and click **Continue**. A summary of the configuration of the imported VM image is displayed. You can review this information then click **Import** without changing anything.

Once the process completes, a new VM called `freeDiameter` appears in the list in the main *VirtualBox* application screen.

3. On the **Oracle VM VirtualBox Manager** screen, select this new machine, and click the **Start** button. A new window appears and displays the screen of your `freeDiameter` VM booting – exactly as if it was the physical display of a separate machine, starting after power on.

Once the boot sequence is complete, the VM will enter the Ubuntu desktop environment.

Optional If you are not using a US keyboard ('QWERTY'), click the **Keyboard Layout** icon located in the upper right of the VM's desktop and select **Keyboard Layout Settings…**. The **Keyboard Layout** screen appears. Press the small + sign at the bottom of the screen to add the appropriate layout option. Use the controls near the + to move this added layout to the top of the list, so that it will be selected automatically when the machine restarts.

4. Read the *freeDiameter-START.pdf* document on the VM's desktop. It contains general information about this VM installation, such as file structure and the username and password. It may also contain pertinent updates since this book was published.

5. To power down the `freeDiameter` VM, click the small gear icon at the upper right of the VM screen, then click the **Shut Down…** option. The VM window will disappear once the shutdown sequence completes.

You are now ready to set up your first example network topology.

A Note about Updating the VM The VM's Internet access is enabled by default to make it easy for you to install additional software packages you may want, such as your favorite text editor.

To minimize the risk of breaking the compatibility with the included `freeDiameter`, be careful with any system updates of the VM. In general, installing the security updates or additional packages should not cause issues, but if you are upgrading the base Ubuntu distribution to a newer release, some scripts included in this VM may not work as expected and `freeDiameter` may need to be updated to support changes in dependent packages.

If you are uncertain about deploying updates, it is safer not to upgrade any package and to avoid using this VM environment to browse the Internet, instead using your host machine for browsing.

It is important to note that the above guidance applies only to your experimental Diameter setups using the provided VM. For operational deployments, you must have a software update process in place to ensure that security vulnerabilities are patched as quickly as possible.

A.6   Creating a Connection Between Two Diameter Peers

Our first example topology consists of two Diameter peers, named client.example.net and server.example.net, that can exchange data over an IP network, as shown in Figure A.1. The clouds conventionally represent an unspecified IP network, and the small planet represents the Internet, although in our VM environment the communication will be contained in virtual network links. We will create this topology step by step in this section.

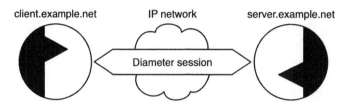

client.example.net   IP network   server.example.net

Diameter session

Figure A.1 Topology for the example setup.

A.6.1 Building client.example.net

We will start by creating the client.example.net peer.

1. In the *VirtualBox* application, ensure that the `freeDiameter` VM is switched off. Note: Do not delete the master `freeDiameter` VM image, or all linked clones will stop functioning.
2. Select the `freeDiameter` VM in the list. From the **Machine** menu, select **Clone**. On the **New machine name** screen, enter 'fD-client.example.net' as the new machine name, and enable the **Reinitialize the MAC address of all network cards** option. This is important, as otherwise no communication between the cloned machines will be possible.
3. Click the **Continue** button. On the **Clone type** screen that appears, select the **Linked clone** option to save significant disk space and time.
4. Click the **Clone** button. A new VM appears in the main *VirtualBox* application.
5. Select this new VM and click the **Start** button. Allow the VM to complete its startup sequence.
6. Open the **Terminal** application by double-clicking the icon located on the desktop. A new terminal window appears and displays a blue dollar sign **$** followed by a blinking cursor; this is called the shell prompt, and the blinking cursor indicates that it is ready to handle a new command that you will type on your keyboard.
   When we describe the commands that must be entered in the terminal we will start the line with a bold dollar sign. Enter the command as given after the bold dollar sign.
7. Apply the network configuration profile to this VM by entering the following command and pressing **Enter**:

```
$ nw_configure.sh client.example.net
```

As the script `nw_configure.sh` executes, the following lines will be displayed:

```
$ nw_configure.sh client.example.net
Tearing down all network interfaces
Applying new configuration
Bringing up the network with the new configuration
networking stop/waiting

Your VM is now configured for client.example.net use-case

$
```

The network has now been configured successfully. You can no longer access the Internet from this VM.

8. To prepare the `freeDiameter` instance that will run on this machine, enter the following command (the parameter starts with the digit 1, not the lower case letter L):

```
$ fD_configure.sh 1_cli
```

The `fD_configure.sh` script builds and configures the `freeDiameter` framework and takes some time to complete, outputting many lines of messages. When the terminal returns to the prompt, a fresh set of `freeDiameter` framework binaries (*libfdproto, libfdcore*, and *freeDiameterd*) can be found in the folder `/home/free-diameter/freeDiameter/test/`, along with configuration files specific to our first test scenario.

The first VM instance, fD-client.example.net, is now ready to use, and we will next prepare the second machine of our topology, server.example.net.

Restoring Network Defaults   You can return to the previous network configuration that enables Internet access by using the command:

```
$ nw_configure.sh default
```

However, you must reapply the client.example.net configuration before continuing with these examples.

A.6.2   Building server.example.net

The creation of this other VM is similar to the creation of the fD-client.example.net VM.

1. In the *VirtualBox* application, ensure that the `freeDiameter` VM is switched off. The fD-client.example.net VM can be left running.
2. Select the `freeDiameter` VM in the list. From the **Machine** menu, select **Clone**. On the **New machine name** screen, enter 'fD-server.example.net' as the new machine name, and enable the **Reinitialize the MAC address of all network cards** option.
3. Click the **Continue** button. On the **Clone type** screen that appears, select the **Linked clone** option.
4. Click the **Clone** button. A new VM appears in the main *VirtualBox* application.
5. Select this new VM and click the **Start** button. Allow the VM to complete its startup sequence.
6. Open a terminal window and run the following command:

```
$ nw_configure.sh server.example.net
```

The test network topology is now configured.
7. Enter the following command to prepare the `freeDiameter` framework on the server:

```
$ fD_configure.sh 1_srv
```

Testing the IP Configuration   You can test the IP configuration by running the following command:

```
$ ping client.example.net
PING client.example.net (192.168.35.5) 56(84) bytes of data.
64 bytes from client.example.net (192.168.35.5): icmp_req=1 ttl=64
   time=0.715 ms
64 bytes from client.example.net (192.168.35.5): icmp_req=2 ttl=64
   time=1.57 ms
```

You will see a new line of information displayed each second. This line shows the duration of a simple round trip exchange of data over the network with the client.example.net machine. If a time value is displayed, like above, the network configuration on both the client and the server are correctly applied for this test. At the bottom of the VM window, the network sign blinks periodically to show network activity on both virtual machines, emphasizing the exchange of data over the virtual network.

To stop the ping command, press Ctrl-C. The terminal will display a summary of the packet exchanges, then will return to the prompt, ready to receive the next command.

A.6.3 Creating the Diameter Connection

1. Ensure both fD-server.example.net and fD-client.example.net VMs are running.
2. Arrange the VM desktop windows so you can see the two VM screens side by side. Maximize the terminal window in each VM so the lines will wrap as little as possible.
3. Enter the following command on the fD-server.example.net VM Terminal prompt:

```
$ freeDiameterd
05:35:17  NOTI    libfdproto '1.2.0-1247(2c09e2545188)' initialized.
05:35:17  NOTI    libgnutls '2.12.14' initialized.
05:35:17  NOTI    libfdcore '1.2.0-1247(2c09e2545188)' initialized.
05:35:18  NOTI    All extensions loaded.
05:35:18  NOTI    freeDiameter configuration:
05:35:18  NOTI       Default trace level .... : +3
05:35:18  NOTI       Configuration file ..... : /home/freediameter/
   freeDiameter/test/freeDiameter.conf
05:35:18  NOTI       Diameter Identity ...... : server.example.net (1:17)
05:35:18  NOTI       Diameter Realm ......... : example.net (1:10)
05:35:18  NOTI       Tc Timer ............... : 30
05:35:18  NOTI       Tw Timer ............... : 30
05:35:18  NOTI       Local port ............. : 3868
05:35:18  NOTI       Local secure port ...... : 5658
05:35:18  NOTI       Number of SCTP streams . : 30
05:35:18  NOTI       Number of clients thr .. : 5
05:35:18  NOTI       Number of app threads .. : 4
05:35:18  NOTI       Local endpoints ........ : Default (use all
   available)
05:35:18  NOTI       Local applications ..... : (none)
05:35:18  NOTI       Flags : - IP ........... : Enabled
05:35:18  NOTI               - IPv6 ......... : Enabled
05:35:18  NOTI               - Relay app .... : Enabled
05:35:18  NOTI               - TCP .......... : Enabled
05:35:18  NOTI               - SCTP ......... : Enabled
05:35:18  NOTI               - Pref. proto .. : SCTP
05:35:18  NOTI               - TLS method ... : Separate port
05:35:18  NOTI       TLS :   - Certificate .. : /home/freediameter/
   freeDiameter/test/cert.pem
05:35:18  NOTI               - Private key .. : /home/freediameter/
   freeDiameter/test/privkey.pem
```

```
05:35:18  NOTI              - CA (trust) ... :
    /home/freediameter/freeDiameter/test/ca.pem (1 certs)
05:35:18  NOTI              - CRL .......... : (none)
05:35:18  NOTI              - Priority ..... : (default: 'NORMAL')
05:35:18  NOTI              - DH bits ...... : 1024
05:35:18  NOTI        Origin-State-Id ........ : 1384245317
05:35:18  NOTI        Loaded extensions: '/home/freediameter/
    freeDiameter/test/lib/dbg_msg_dumps.fdx'[0x2020], loaded
05:35:18  NOTI        Local server address(es): 192.168.35.10{---L-}
05:35:18  NOTI        freeDiameterd daemon initialized.
05:35:18  NOTI        CONNECT FAILED to client.example.net: All
    connection attempts failed, will retry later
```

These lines are produced by the `freeDiameter` framework. Each line starts with a timestamp, followed by an indicator of the importance of the line (NOTI for notifications, ERROR for errors, etc.) and the message itself. The above capture shows the initialization sequence of `freeDiameter`, starting with the different component versions displayed, followed by a summary of the configuration, and finally a notification that the initialization is complete, which means the framework is ready to handle Diameter connections.

This machine is configured to establish a Diameter connection with client.example .net, so it attempts to establish a connection to that machine periodically. The attempts are failing because there is no Diameter stack running on the other machine to accept the incoming connection.

4. Run the `freeDiameterd` command in the fD-client.example.net VM window:

```
$ freeDiameterd
05:36:30  NOTI  libfdproto '1.2.0-1247(2c09e2545188)' initialized.
05:36:30  NOTI  libgnutls '2.12.14' initialized.
05:36:30  NOTI  libfdcore '1.2.0-1247(2c09e2545188)' initialized.
05:36:32  NOTI  All extensions loaded.
05:36:32  NOTI  freeDiameter configuration:
05:36:32  NOTI     Default trace level .... : +3
05:36:32  NOTI     Configuration file ..... : /home/freediameter/
    freeDiameter/test/freeDiameter.conf
05:36:32  NOTI     Diameter Identity ...... : client.example.net (1:17)
05:36:32  NOTI     Diameter Realm ......... : example.net (1:10)
05:36:32  NOTI     Tc Timer ............... : 30
05:36:32  NOTI     Tw Timer ............... : 30
05:36:32  NOTI     Local port ............. : 3868
05:36:32  NOTI     Local secure port ...... : 5658
05:36:32  NOTI     Number of SCTP streams . : 30
05:36:32  NOTI     Number of clients thr .. : 5
05:36:32  NOTI     Number of app threads .. : 4
05:36:32  NOTI     Local endpoints ........ : Default (use all
    available)
05:36:32  NOTI     Local applications ..... : (none)
05:36:32  NOTI     Flags : - IP .......... : Enabled
05:36:32  NOTI             - IPv6 ......... : Enabled
05:36:32  NOTI             - Relay app .... : Enabled
05:36:32  NOTI             - TCP .......... : Enabled
05:36:32  NOTI             - SCTP ......... : Enabled
05:36:32  NOTI             - Pref. proto .. : SCTP
05:36:32  NOTI             - TLS method ... : Separate port
05:36:32  NOTI     TLS :   - Certificate .. : /home/freediameter/
    freeDiameter/test/cert.pem
```

```
05:36:32  NOTI               - Private key .. :
    /home/freediameter/freeDiameter/test/privkey.pem
05:36:32  NOTI               - CA (trust) ... :
    /home/freediameter/freeDiameter/test/ca.pem (1 certs)
05:36:32  NOTI               - CRL ......... : (none)
05:36:32  NOTI               - Priority ..... : (default: 'NORMAL')
05:36:32  NOTI               - DH bits ...... : 1024
05:36:32  NOTI       Origin-State-Id ........ : 1384245390
05:36:32  NOTI    Loaded extensions: '/home/freediameter/
    freeDiameter/test/lib/dbg_msg_dumps.fdx'[0x2020], loaded
05:36:32  NOTI    Local server address(es): 192.168.35.5{---L-}
05:36:32  NOTI    freeDiameterd daemon initialized.
05:36:32  NOTI    SND to 'server.example.net':
    'Capabilities-Exchange-Request'0/257 f:R--- src:'(nil)' len:152
05:36:32  NOTI    RCV from 'server.example.net': (no model)0/257 f:----
    src:'server.example.net' len:164
05:36:32  NOTI    CONNECTED TO 'server.example.net' (SCTP,TLS,soc#17)
05:36:32  NOTI    'STATE_WAITCEA'        -> 'STATE_OPEN'
    'server.example.net'
```

Note that the following additional lines have appeared in parallel on the fD-server
.example.net screen:

```
05:36:32  NOTI    RCV from '<unknown peer>': (no model)0/257 f:R---
    src:'(nil)' len:152
05:36:32  NOTI    CONNECTED TO 'client.example.net' (SCTP,TLS,soc#16)
05:36:32  NOTI    SND to 'client.example.net': 'Capabilities-
    Exchange-Answer'0/257 f:---- src:'(nil)' len:164
05:36:32  NOTI    'STATE_CLOSED' -> 'STATE_OPEN' 'client.example.net'
```

Here is what has happened: after the initialization of `freeDiameter` on the
fD-client.example.net VM, the client attempted to connect to server.example.net and
succeeded. The client sent a Diameter message called Capabilities-Exchange-Request
(line starting with SND) to the server, where it appears as a received (RCV) message.
The server sends the corresponding answer, Capabilities-Exchange-Answer, and
each peer displays the information that it is now CONNECTED TO the other peer,
and that the connection is moved to the STATE_OPEN state. From this point, the
peers are able to exchange Diameter traffic. If a peer does not receive a message
on the connection during a configured period (30 seconds by default), it sends a
message called Device-Watchdog-Request to verify that the connection is still alive.
You can observe this if you wait for a few seconds:

```
11:28:42  NOTI    SND to 'server.example.net': 'Device-Watchdog-
    Request'0/280 f:R--- src:'(nil)' len:80
11:28:42  NOTI    RCV from 'server.example.net': (no model)0/280 f
    :---- src:'server.example.net' len:92
```

5. To stop the `freeDiameter` instance running on the fD-client.example.net VM,
   ensure that its terminal window has the focus, then enter Ctrl-C.

```
^C (This symbol on the left represents the Ctrl-C key sequence)
11:28:48  FATAL! Initiating freeDiameter shutdown sequence (3)
11:28:48  NOTI    freeDiameterd framework is stopping...
11:28:48  NOTI    Shutting down server sockets...
11:28:48  NOTI    Sending terminate signal to all peer connections
11:28:48  NOTI    'STATE_OPEN'     -> 'STATE_CLOSING_GRACE'
    'server.example.net'
```

```
11:28:48  NOTI    SND to 'server.example.net': 'Disconnect-Peer-
    Request'0/282 f:R--- src:'(nil)' len:80
11:28:48  NOTI    Waiting for connections shutdown... (16 sec max)
11:28:48  NOTI    RCV from 'server.example.net': (no model)0/282 f
    :---- src:'server.example.net' len:80
11:28:48  NOTI    server.example.net: Going to ZOMBIE state (no more
    activity)
11:28:48  ERROR   ERROR: in '(fd_tls_rcvthr_core(conn, conn->
    cc_tls_para.session))' :         Transport endpoint is not connected
11:28:48  NOTI    'STATE_CLOSED' -> STATE_ZOMBIE (terminated)
    'server.example.net'
$
```

The above log shows the shutdown sequence of `freeDiameter` initiated when we press Ctrl-C. We can see that the connection is not closed immediately, but a final Disconnect-Peer-Request message is sent to the other peer, and connection is kept open waiting for the corresponding answer. This final exchange is also part of the Diameter base protocol and allows the remote peer some control over the connection termination, for example by sending pending messages. After the answer has been received, `freeDiameter` terminates and the Terminal prompt reappears.

On the other VM screen, you can see that new periodical attempts are made to re-establish the connection.

6. You can now stop the `freeDiameter` instance running on the fD-server.example .net VM by ensuring that its terminal window has the focus, then entering Ctrl-C.

This concludes our first experiment, in which we have configured two VMs and let them establish a Diameter link using the Diameter Base protocol. We have seen three different Diameter commands in action: Capabilities-Exchange at the initial connection establishment, Device-Watchdog exchanged periodically over the open connection, and Disconnect-Peer when the connection is being closed by one of the peers.

Appendix B

`freeDiameter` from Sources

B.1 Introduction

This appendix provides more details on how to build `freeDiameter`. It is recommended for anyone interested in developing new features in Diameter or hacking in the existing `freeDiameter` code. As a prerequisite, the reader should be familiar with software development and the Linux environment.

The general steps to compile and then run `freeDiameter` are the following:

1. Install the required tools and dependencies.
2. Obtain the source code.
3. Set up your build environment.
4. Run **make**, which will in turn call the C compiler and other tools needed to compile `freeDiameter`.
5. Create a `freeDiameter` configuration file.
6. Run and test `freeDiameter`.

The following sections provide more details on these steps, which will create the environment covered in Appendix A.

B.2 Tools and Dependencies

As `freeDiameter` source code is not tied to any integrated development environment, almost any system can be used as a development machine. There are, however, a few dependencies that need to be matched. Use the latest stable release for your operating system when installing or updating dependencies.

Operating System
The `freeDiameter` code follows the programming interface specified by POSIX, so it can run on any UNIX® compliant system, and it has been tested successfully on several flavors of GNU/Linux, FreeBSD, and Mac OS X. See http://www.unix.org for details on the POSIX API.

Diameter: New Generation AAA Protocol – Design, Practice, and Applications, First Edition.
Hannes Tschofenig, Sébastien Decugis, Jean Mahoney and Jouni Korhonen.
© 2019 John Wiley & Sons Ltd. Published 2019 by John Wiley & Sons Ltd.

Build System

`freeDiameter` uses **CMake** for configuring its build system. **CMake** is a command-line tool available on most platforms and usually comes with a graphical user interface, **ccmake**, for easier use. For more information about **CMake** and **ccmake**, see http://cmake.org.

The **make** tool is used for automating the build. `freeDiameter` has been tested with **GNU Make**, the default version in many distributions.

Compilers

The source code of `freeDiameter` is written mainly in C language, but Lex and Yacc languages are also used, and some extensions use additional languages. In order to compile the C source code, you will need a C/C++ compiler. The default GNU C compiler **gcc toolchain** is a good choice, but in some cases a different compiler may be preferred; such an option can be easily configured in CMake.

You will need the **GNU Flex** and **GNU Bison** tools to compile the Lex and Yacc files – tokenizer and grammar parser, respectively – used for parsing the `freeDiameter` configuration file.

Source Code Management

The source code of `freeDiameter` is managed with Mercurial, an open-source and platform-independent source code management (SCM) tool. Although not absolutely necessary for retrieving the source code, Mercurial allows access to any previous version of the `freeDiameter` project and will allow you to check code into the `freeDiameter` repository if you choose to contribute. To obtain Mercurial, visit https://www.mercurial-scm.org/.

B.2.1 Runtime Dependencies

`freeDiameter` depends on a number of libraries used at runtime and their corresponding header files for compilation.

B.2.1.1 SCTP

At the time of writing, the SCTP protocol is not fully supported by default on many systems, and requires a library to be installed, e.g., *libsctp1* and *libsctp-dev* packages on Ubuntu. The SCTP library for OS X, *libusrsctp*, is not equivalent to other SCTP libraries, so you will need to disable support for SCTP in order to build `freeDiameter` in OS X. This is described in the section that covers the build process.

B.2.1.2 TLS

`freeDiameter` depends on the **GnuTLS** library for the support of TLS security. To download this library, visit http://gnutls.org.

`freeDiameter` also has a dependency on *libgcrypt*, a general-purpose cryptographic library. Visit https://www.gnu.org/software/libgcrypt/ for pointers on where to download it.

B.2.1.3 Internationalized Domain Names

For the support of internationalized domain names, the **GNU IDN** library is used and can be downloaded from http://www.gnu.org/software/libidn. However, if this feature is not required in the target deployment, a **CMake** setting, described below, can toggle the dependency off.

B.3 Obtaining `freeDiameter` Source Code

The source code of `freeDiameter` is freely accessible from the project website: http://www.freediameter.net/. Using Mercurial, clone the repository from the project website to a local directory:

1. Create a directory called `freeDiameter`.
2. Change to the `freeDiameter` directory.
3. Type at the command line:

```
$ hg clone http://www.freediameter.net/hg/freeDiameter src
```

This copies `freeDiameter` source files into a directory named `src` and duplicates the complete history of the `freeDiameter` project locally, making it easy to access any version of the project. To list versions and check out a specific one:

1. Change to the `src` directory.
2. Type at the command line:

```
$ hg tags
```

A list of versions is presented:

```
tip             1315:8662db9f6105
1.2.1           1314:2cb8d71a405d
1.2.1b          1312:6446c0eea547
1.2.1a          1310:9caedf4a058b
```

3. To check out an older version, type at the command line:

```
$ hg checkout 1.2.1
```

4. To download an archive of a specific version of the source code, use the following pattern for the URL, replacing `N.N.N` with any released version corresponding to the tags in Mercurial:

```
$ wget http://www.freediameter.net/hg/freeDiameter/archive/N.N.N.tar.gz
```

The complete list of tags can be found at http://www.freediameter.net/hg/freeDiameter/tags.

To read the most up-to-date information about dependencies and how to further configure your particular environment, look at the *INSTALL* files found in the `src` directory.

B.4 Configuring the Build

Disabling SCTP Support

If you do not want to support SCTP, or your system cannot support SCTP, build `free-Diameter` following these steps:

1. Change to the `freeDiameter` directory.
2. Create a `build` directory.
3. Change to the `build` directory.
4. Type at the prompt:

```
$ cmake ../src -DDISABLE_SCTP:BOOL=ON
```

This populates the `build` directory with subdirectories and files for the build. Once you have run **CMake** to disable SCTP, you may run **CMake** again, or **ccmake**, to make any other build configuration changes.

Interactive Mode for Configuration

`freeDiameter` offers flexible build configuration with **ccmake**, an interactive interface to **CMake** for configuring the build. To use this interactive interface:

1. Change to the `freeDiameter` directory.
2. Create a `build` directory if you have not already done so.
3. Change to the `build` directory.
4. Type at the prompt:

```
$ ccmake ../src
```

The interactive interface will be displayed. Navigation guidance can be found at the bottom of the screen:

```
                             Page 0 of 1
   EMPTY CACHE



EMPTY CACHE:
Press [enter] to edit option               CMake Version 3.7.0
Press [c] to configure
Press [h] for help          Press [q] to quit without generating
Press [t] to toggle advanced mode (Currently Off)
```

5. Press the c key (configure) to list all `freeDiameter` variables that **ccmake** needs to configure. If you are missing any dependencies, errors will be displayed, but **ccmake** will continue on to the variables list similar to the following:

```
                                           Page 1 of 2
   ALL_EXTENSIONS               *OFF
   BUILD_ACL_WL                 *ON
   BUILD_APP_ACCT               *OFF
   BUILD_APP_DIAMEAP            *OFF
   BUILD_DBG_INTERACTIVE        *OFF
   BUILD_DBG_MONITOR            *ON
   BUILD_DICT_DCCA              *ON
```

```
BUILD_DICT_EAP                    *ON
BUILD_DICT_LEGACY_XML             *OFF
BUILD_RT_LOAD_BALANCE             *ON

ALL_EXTENSIONS: Build ALL available extensions? (disable to select
    individual components)
Press [enter] to edit option                    CMake Version 3.7.0
Press [c] to configure
Press [h] for help         Press [q] to quit without generating
Press [t] to toggle advanced mode (Currently Off)
```

6. Press the t key (toggle) to access a full list of the configuration options. By default many options are hidden since they usually do not need to be changed.
7. Use arrow keys to navigate the list and press **Enter** to edit the selected option.
8. Once you have configured your options, type c again to store this configuration and reiterate the process, as setting some options may have unlocked new features to be configured. That is, when enabling some extensions, new dependencies will need to be resolved.
9. When finished, press the g (generate) key and **ccmake** will write the Makefiles that are needed for the next step. In addition, **ccmake** generates the *freeDiameter-host.h* file in the include/freeDiameter/ directory that contains values that will be used by the freeDiameter source code, such as system parameters and features.

The following describes a few of the useful options. Refer to the embedded help text for additional information. In general all options prefixed with CMAKE_* are standard options that are documented in the **CMake** product:

ALL_EXTENSIONS Setting this to ON will enable the compilation of all the provided extensions. By setting this option to OFF, you can select which extensions are built by configuring the BUILD_* options. For example, set the BUILD_APP_DIAMEAP option to ON to compile the Diameter EAP server application (the *app_diameap.fdx* extension).

CMAKE_BUILD_TYPE This configures a set of flags for different profiles. When developing or testing freeDiameter, set this to Debug, which unlocks additional options (the **CMake** configuration process is iterative).

CMAKE_C_COMPILER Change here if you want to use a different C compiler.

CMAKE_INSTALL_PREFIX If the make install command is used, this configuration controls where the files must be installed. See also the INSTALL_*_SUFFIX options described in the installation section below.

DEBUG_WITH_META Set to ON if you want logs with additional contextual information such as the thread ID, the producing function, etc.

DEBUG_SCTP Set to ON for more verbose messages in the SCTP code of freeDiameter. Only useful when working on this part of the code.

SCTP_USE_MAPPED_ADDRESSES Some earlier versions of the SCTP stack did not handle mixed IP and IPv6 arrays, so setting to ON will force the use of IPv4-mapped IPv6 addresses [1] to work around the issue.

FLEX_EXECUTABLE This specifies the flex tool that will be used to compile the Lex files.

GNUTLS_INCLUDE_DIR, GNUTLS_LIBRARY Specify the GnuTLS header files and library to use for compiling freeDiameter. Note that the library used at runtime is selected using the default loader mechanism, which can be controlled with the `LD_LIBRARY_PATH` environment variable. These options are useful when working with a non-default GnuTLS library.

DIAMID_IDNA_REJECT Set to `OFF` if you do not have the GNU *libidn* library installed or internationalized domain names are not required in the target deployment.

Instead of using the **ccmake** interactive process, you can use the **CMake** command line directly:

```
$ cmake -DDISABLE_SCTP:BOOL=ON -DDEBUG_WITH_META:BOOL=ON /path/to/src
```

B.5 Compiling `freeDiameter`

Once the *Makefile* has been generated as described in the previous section, type the following to compile the source code:

```
$ make
```

If the compiler is auto-configured properly, you should see an output similar to the following:

```
[  1%] Retrieving version of the hg repository
-- Source version: 1286(ecb844d6d87d)
[  1%] Built target version_information
Scanning dependencies of target libfdproto
[  2%] Building C object libfdproto/CMakeFiles/libfdproto.dir/
    dictionary.c.o
[  3%] Building C object libfdproto/CMakeFiles/libfdproto.dir/
    dictionary_functions.c.o
[  4%] Building C object libfdproto/CMakeFiles/libfdproto.dir/dispatch.c.o
[  5%] Building C object libfdproto/CMakeFiles/libfdproto.dir/fifo.c.o
[  6%] Building C object libfdproto/CMakeFiles/libfdproto.dir/init.c.o
[  7%] Building C object libfdproto/CMakeFiles/libfdproto.dir/lists.c.o
[  8%] Building C object libfdproto/CMakeFiles/libfdproto.dir/log.c.o
...
Scanning dependencies of target dbg_msg_dumps
[ 98%] Building C object extensions/dbg_msg_dumps/CMakeFiles/
    dbg_msg_dumps.dir/dbg_msg_dumps.c.o
Linking C shared module ../dbg_msg_dumps.fdx
[ 98%] Built target dbg_msg_dumps
Scanning dependencies of target dbg_rt
[100%] Building C object extensions/dbg_rt/CMakeFiles/dbg_rt.dir/dbg_rt
    .c.o
Linking C shared module ../dbg_rt.fdx
[100%] Built target dbg_rt
```

By default, **CMake** generates silent Makefile rules, but the compilation line can be made visible by typing:

```
$ make VERBOSE=1
```

B.6 Installing `freeDiameter`

For this tutorial, we will not use the installation mechanism that is described below, but will run `freeDiameter` from the build tree directly, which is the standard method during development. Note that the information below is provided for completeness only.

To install from the source, specify the top-level directory where `freeDiameter` will be installed with the **CMake** directive, `CMAKE_INSTALL_PREFIX`, which is `/usr/local` on most systems by default, and specify the subdirectory for different types of built files with the `INSTALL_*_SUFFIX` directives. For example, the following directives will install extensions into `/usr/local/lib/freeDiameter/`:

- `CMAKE_INSTALL_PREFIX=/usr/local`
- `INSTALL_EXTENSIONS_SUFFIX=lib/freeDiameter`

After **CMake** has generated a *Makefile* with the proper path set, run the following command to copy the generated files into the configured locations. If the target is a system location, then run this command as a privileged user:

```
$ make install
```

Note that there is no `make uninstall` target generated by default, but a list of the installed files can be found in *install_manifest.txt*. If you want to remove all the files that have been installed by the `make install` command, use the following command:

```
$ xargs rm < install_manifest.txt
```

B.7 `freeDiameter` Configuration File

`freeDiameter` reads its configuration during startup from a file specified on the command line. This configuration file contains the settings for the core module and references to additional files that are related to specific extensions. The *freeDiameter.conf.sample* file in the `/src/doc/` directory documents all the parameters. The parameters in this file are grouped as follows:

Network protocol IP and IPv6 related configuration.

Transport protocol Information such as the port numbers to use, the number of SCTP streams, etc.

Transport layer security Certificate, private key, trusted authorities.

Diameter protocol Identity, timers values, number of application threads.

Peers List of the Diameter peers to maintain a connection to. Extension may add additional peers dynamically.

Extensions The `freeDiameter` extensions to be loaded, with their respective configuration files when appropriate. Documentation for the extensions configuration files is found in the `src/doc/` directory, and named after the extension.

Most parameters have reasonable defaults. For example, `freeDiameter` will use the hostname of the machine as DiameterIdentity unless the configuration file contains an `Identity=...` line. Only the information on the private key and certificate pertaining to the TLS configuration are mandatory parameters that must be present in the file because no default can be set. An example of a working configuration file is *freeDiameter-1.conf* found in the `/src/doc/single_host/` directory.

B.8 Running and Debugging `freeDiameter`

We are now going to set two Diameter peers running on a single machine, and let them exchange some example messages:

1. Change to the `build` directory.
2. Configure the *Makefile* to compile the *test_app.fdx* extension:

   ```
   $ cmake -DBUILD_TEST_APP:BOOL=ON ../src
   ```

3. Run **make**:

   ```
   $ make
   ```

4. Copy configuration files for running two peers:

   ```
   $ cp ../src/doc/single_host/* .
   ```

5. Use OpenSSL to generate a local certificate infrastructure for this test that matches the identities declared in the configuration files:

   ```
   $ bash make_certs.sh
   ```

   We now have two sets of files in the current directory for two peers. The first set contains *freeDiameter-1.conf*, the main configuration for a peer with identity peer1.localdomain. This file references *cacert.pem*, the certificate authority for the local test, and *peer1.cert.pem* and *peer1.key.pem*, respectively the certificate and private key for this peer1. The configuration file also references *test_app1.conf*, the configuration file for the *test_app.fdx* extension for peer1. The second set contains *freeDiameter-2.conf*, with a similar structure as peer1. Note that only the certificate authority file is shared between both peers, since this is the common root of trust.

6. Start the stack. The `-c` flag allows you to specify the configuration file:

   ```
   $ freeDiameterd/freeDiameterd -c freeDiameter-1.conf
   ```

Provided that everything was done properly and you have all the dependencies available, you will see the following output:

```
16:48:20  NOTI    libfdproto '1.2.1-1291(0fa8207cc91a)' initialized.
16:48:20  NOTI    libgnutls '3.3.14' initialized.
16:48:20  NOTI    libfdcore '1.2.1-1291(0fa8207cc91a)' initialized.
16:48:20  NOTI    Extension Test_App initialized with configuration: '
    test_app1.conf'
16:48:20  NOTI    All extensions loaded.
16:48:20  NOTI    freeDiameter configuration:
16:48:20  NOTI      Default trace level .... : +3
16:48:20  NOTI      Configuration file ..... : freeDiameter-1.conf
16:48:20  NOTI      Diameter Identity ...... : peer1.localdomain (1:17)
16:48:20  NOTI      Diameter Realm ......... : localdomain (1:11)
16:48:20  NOTI      Tc Timer ............... : 30
```

```
16:48:20  NOTI      Tw Timer ............... : 30
16:48:20  NOTI      Local port ............. : 3868
16:48:20  NOTI      Local secure port ...... : 5658
16:48:20  NOTI      Number of SCTP streams . : 30
16:48:20  NOTI      Number of clients thr .. : 5
16:48:20  NOTI      Number of app threads .. : 4
16:48:20  NOTI      Local endpoints ........ : Default (use all
   available)
16:48:20  NOTI      Local applications ..... : App: 16777215,Au--,Vnd
   :999999
16:48:20  NOTI      Flags : - IP ........... : Enabled
16:48:20  NOTI              - IPv6 ......... : Enabled
16:48:20  NOTI              - Relay app .... : Enabled
16:48:20  NOTI              - TCP .......... : Enabled
16:48:20  NOTI              - SCTP ......... : DISABLED (at compilation)
16:48:20  NOTI              - Pref. proto .. : SCTP
16:48:20  NOTI              - TLS method ... : Separate port
16:48:20  NOTI      TLS :   - Certificate .. : peer1.cert.pem
16:48:20  NOTI              - Private key .. : peer1.key.pem
16:48:20  NOTI              - CA (trust) ... : cacert.pem (1 certs)
16:48:20  NOTI              - CRL .......... : (none)
16:48:20  NOTI              - Priority ..... : (default: 'NORMAL')
16:48:20  NOTI              - DH bits ...... : 1024
16:48:20  NOTI      Origin-State-Id ........ : 1434790100
16:48:20  NOTI      Loaded extensions: 'extensions/test_app.fdx'[test_app1
   .conf], loaded
16:48:20  NOTI      {signal:30}'test_app.bench'->0x10d3d3120
16:48:20  NOTI      Local server address(es): 192.168.1.6{---L-}
16:48:20  NOTI      freeDiameterd daemon initialized.
```

In order to run our two-peers scenario, we need to launch the second peer.

1. In a separate terminal window, launch the second peer using the configuration *freeDiameter-2.conf*:

```
$ freeDiameterd/freeDiameterd -c freeDiameter-2.conf
```

Once both instances are running, the following output shows that a connection was successfully established:

```
17:05:02  NOTI    Connected to 'peer2.localdomain' (TCP,soc#9),
    remote capabilities:
17:05:02  NOTI         Capabilities-Exchange-Request(257)[R---],
    Length=208, Hop-By-Hop-Id=0x2ac1d6b7, End-to-End=0x0bef23e1, {
    Origin-Host(264)[-M]="peer2.localdomain" }, { Origin-Realm
    (296)[-M]="localdomain" }, {  Origin-State-Id(278)[-M
    ]=1445159102 (0x562360be) }, { Host-IP-Address(257)[-M]
    =192.168.1.6 }, { Vendor-Id(266)[-M]=0 (0x0) }, { Product-
    Name(269)[--]="freeDiameter" }, { Firmware-Revision(267)
    [--]=10201 (0x27d9) }, { Inband-Security-Id(299)[-M]='
    NO_INBAND_SECURITY' (0 (0x0)) }, { Vendor-Specific-Application
    -Id(260)[-M]={ Auth-Application-Id(258)[-M]=16777215 (0xfffffff
    ) }, { Vendor-Id(266)[-M]=999999 (0xf423f) } }, { Auth-
    Application-Id(258)[-M]=4294967295 (0xffffffff) }, { Supported
    -Vendor-Id(265)[-M]=999999 (0xf423f) }
17:05:02  NOTI    No TLS protection negotiated with peer 'peer2.
localdomain'.
17:05:02  NOTI    'STATE_CLOSED' -> 'STATE_OPEN' 'peer2.localdomain'
```

2. Trigger a message exchange with the following command in another terminal window:

```
$ killall -USR1 freeDiameterd
```

Once triggered with this signal, both peers instances generate a test message and send to their peer, as shown on the console of the running processes:

```
SEND 71f32454 to 'localdomain' (-)
ECHO Test-Request received from 'peer2.localdomain', replying...
RECV 71f32454 (Ok) Status: 2001 From 'peer2.localdomain' ('
    localdomain') in 0.001083 sec
```

3. When done, press Ctrl-C in each terminal window to halt the peers.

This concludes our tutorial for running `freeDiameter` from the source code. The remainder of this chapter provides more information on debugging, hacking, and extending the framework.

B.9 Extensions for Debug Support

Because `freeDiameter` was initially designed for academics and prototyping, it comes with a fair number of features built for facilitating the modification of the code.

B.9.1 Extended Trace

When you are developing and debugging, you may need to have more verbose output of the stack. There are two debugging options that you can apply.

First, in the **CMake** step, set the `DEBUG_WITH_META` option to `ON`. The `DEBUG_WITH_META` option adds a more precise timestamp, the name of the thread that triggered a log – useful since `freeDiameter` is multithreaded – and the function name, source file, and line where the log was triggered.

The second debugging option is to increase the verbosity level with the `-dd` switch when you start `freeDiameter`:

```
$ freeDiameterd/freeDiameterd -c freeDiameter-1.conf -dd
```

You will now have much more information available on what is happening in the stack:

```
06/20/15,16:55:56.298326  NOTI   pid:Main in fd_core_initialize@log.c
    :188: libfdproto '1.2.1-1296(f02561ecd19a)' initialized.
06/20/15,16:55:56.298882  NOTI   pid:Main in fd_core_initialize@log.c
    :201: libgnutls '3.3.14' initialized.
06/20/15,16:55:56.301201  DBG    pid:Main in core_state_set@p_expiry.c
    :78: Core state: 0 -> 1
06/20/15,16:55:56.301223  NOTI   pid:Main in
    fd_core_initialize@p_expiry.c:223: libfdcore '1.2.1-1296(
    f02561ecd19a)' initialized.
06/20/15,16:55:56.304083  DBG    pid:Main in fd_conf_parse@p_psm.c:588:
    Generating fresh Diffie-Hellman parameters of size 1024 (this takes
    some time)...
06/20/15,16:55:56.337098  DBG    pid:Main in
    fd_ext_load@routing_dispatch.c:160: Loading : extensions/test_app.
    fdx
```

```
06/20/15,16:55:56.337916  NOTI   pid:Main in ta_entry@ta_conf.y:170:
   Extension Test_App initialized with configuration: 'test_app1.conf'
06/20/15,16:55:56.337938  DBG   ------- app_test configuration dump:
   ---------
06/20/15,16:55:56.337944  DBG   Vendor Id .......... : 999999
```

On the downside, the readability of the log decreases when these options are used; in addition, if you are storing the output on disk, the file size will increase rapidly. These options should be used only during the development phase.

The -dd command line switch triggers more traces to be displayed. The verbosity can be increased by adding the flag multiple times (e.g., -ddd), but the stack will not run properly if too much time is spent producing the traces. You can limit higher verbosity to a source file or a single function only, which may help while developing or debugging a specific module. Here is an example output:

```
$ ./freeDiameterd/freeDiameterd -c freeDiameter-1.conf --dbg_file
   config.c
10/18/15,02:09:10.010625  NOTI   pid:Main in fd_core_initialize@core.c
   :188: libfdproto '1.2.1-1303(090390e89b1d)' initialized.
10/18/15,02:09:10.015356  NOTI   pid:Main in fd_core_initialize@core.c
   :201: libgnutls '2.12.14' initialized.
10/18/15,02:09:10.015668  DBG   pid:Main in fd_conf_init@config.c:51:
   [DBG_MATCH] [enter] fd_conf_init() {}
10/18/15,02:09:10.015846  DBG   pid:Main in fd_conf_init@config.c:71:
   [DBG_MATCH] Check: ((fd_dict_init(&fd_g_config->cnf_dict)))
10/18/15,02:09:10.016093  DBG   pid:Main in fd_conf_init@config.c:72:
   [DBG_MATCH] Check: ((fd_fifo_new(&fd_g_config->cnf_main_ev, 0)))
10/18/15,02:09:10.016281  DBG   pid:Main in fd_conf_init@config.c:75:
   [DBG_MATCH] Check: gnutls_certificate_allocate_credentials
   (&fd_g_config->cnf_sec_data.credentials)
10/18/15,02:09:10.016504  DBG   pid:Main in fd_conf_init@config.c:76:
   [DBG_MATCH] Check: gnutls_dh_params_init
   (&fd_g_config->cnf_sec_data.dh_cache)
```

B.9.2   Logging Diameter Messages: *dbg_msg_dumps.fdx*

It is useful to get a view of all the Diameter messages going in and out of the framework to understand why an application misbehaves. Although network analyzers such as **Wireshark** are excellent tools for such a purpose, they have some limitations if the traffic is encrypted with TLS or generated locally between two processes instead of sent over a real network. For this reason, `freeDiameter` comes with a Diameter message logger extension, *dbg_msg_dumps.fdx*.

To use the *dbg_msg_dumps.fdx* extension:

1. Enable BUILD_DBG_MSG_DUMPS in your **CMake** configuration:

   ```
   $ cmake -BUILD_DBG_MSG_DUMPS:BOOL=ON ../src
   ```

2. Run **make**:

   ```
   $ make
   ```

3. Include the following line in the *freeDiameter-1.conf* file:

   ```
   LoadExtension = "extensions/dbg_msg_dumps.fdx" : "0x0080";
   ```

4. Start the same test as in previous section:

```
$ freeDiameterd/freeDiameterd -c freeDiameter-1.conf
```

5. In a separate terminal window, launch the second peer:

```
$ freeDiameterd/freeDiameterd -c freeDiameter-2.conf
```

The triggered exchange is displayed as follows for peer1:

```
SEND 643c9869 to 'localdomain' (-)
17:43:33  NOTI    SND to 'peer2.localdomain':
17:43:33  NOTI       'Test-Request'
17:43:33  NOTI          Version: 0x01
17:43:33  NOTI          Length: 152
17:43:33  NOTI          Flags: 0xC0 (RP--)
17:43:33  NOTI          Command Code: 16777214
17:43:33  NOTI          ApplicationId: 16777215
17:43:33  NOTI          Hop-by-Hop Identifier: 0x72309A66
17:43:33  NOTI          End-to-End Identifier: 0x9BA24188
17:43:33  NOTI             {internal data}: src:(nil)(0) rwb:0x0 rt:0
   cb:0x1046103a0,0x0(0x7fbf82e00130) qry:0x0 asso:0 sess:0
   x7fbf82e00c30
17:43:33  NOTI             AVP: 'Session-Id'(263) l=47 f=-M val="
   peer1.localdomain;1445161402;1;app_test"
17:43:33  NOTI             AVP: 'Destination-Realm'(283) l=19 f=-M
   val="localdomain"
17:43:33  NOTI             AVP: 'Origin-Host'(264) l=25 f=-M
   val="peer1.localdomain"
17:43:33  NOTI             AVP: 'Origin-Realm'(296) l=19 f=-M val="
   localdomain"
17:43:33  NOTI             AVP: 'Test-AVP'(16777215) vend='app_test
   vendor'(999999) l=16 f=V- val=1681692777 (0x643c9869)
17:43:33  NOTI    RCV from 'peer2.localdomain':
17:43:33  NOTI       'Test-Request'
17:43:33  NOTI          Version: 0x01
17:43:33  NOTI          Length: 152
17:43:33  NOTI          Flags: 0xC0 (RP--)
17:43:33  NOTI          Command Code: 16777214
17:43:33  NOTI          ApplicationId: 16777215
17:43:33  NOTI          Hop-by-Hop Identifier: 0x2AC1D6DD
17:43:33  NOTI          End-to-End Identifier: 0x0BEF2407
17:43:33  NOTI             {internal data}: src:peer2.localdomain(17)
   rwb:0x0 rt:0 cb:0x0,0x0(0x0) qry:0x0 asso:0 sess:0x0
17:43:33  NOTI             AVP: 'Session-Id'(263) l=47 f=-M val="
   peer2.localdomain;1445159102;2;app_test"
17:43:33  NOTI             AVP: 'Destination-Realm'(283) l=19 f=-M
   val="localdomain"
17:43:33  NOTI             AVP: 'Origin-Host'(264) l=25 f=-M val="
   peer2.localdomain"
17:43:33  NOTI             AVP: 'Origin-Realm'(296) l=19 f=-M val="
   localdomain"
17:43:33  NOTI             AVP: 'Test-AVP'(16777215) vend='app_test
   vendor'(999999) l=16 f=V- val=438792350 (0x1a27709e)
17:43:33  NOTI    RCV from 'peer2.localdomain':
17:43:33  NOTI       'Test-Answer'
17:43:33  NOTI          Version: 0x01
17:43:33  NOTI          Length: 144
17:43:33  NOTI          Flags: 0x40 (-P--)
17:43:33  NOTI          Command Code: 16777214
```

```
17:43:33  NOTI          ApplicationId: 16777215
17:43:33  NOTI          Hop-by-Hop Identifier: 0x72309A66
17:43:33  NOTI          End-to-End Identifier: 0x9BA24188
17:43:33  NOTI             {internal data}: src:peer2.localdomain(17)
   rwb:0x0 rt:0 cb:0x0,0x0(0x0) qry:0x7fbf82e00ae0 asso:0 sess:0x0
17:43:33  NOTI          AVP: 'Session-Id'(263) l=47 f=-M val="peer1.
   localdomain;1445161402;1;app_test"
17:43:33  NOTI          AVP: 'Test-AVP'(16777215) vend='app_test
   vendor'(999999) l=16 f=V- val=1681692777 (0x643c9869)
17:43:33  NOTI          AVP: 'Origin-Host'(264) l=25 f=-M val="
   peer2.localdomain"
17:43:33  NOTI          AVP: 'Origin-Realm'(296) l=19 f=-M val="
   localdomain"
17:43:33  NOTI          AVP: 'Result-Code'(268) l=12 f=-M val='
   DIAMETER_SUCCESS' (2001 (0x7d1))
ECHO Test-Request received from 'peer2.localdomain', replying...
RECV 643c9869 (Ok) Status: 2001 From 'peer2.localdomain' ('
   localdomain') in 0.001339 sec
17:43:33  NOTI   SND to 'peer2.localdomain':
17:43:33  NOTI          'Test-Answer'
17:43:33  NOTI          Version: 0x01
17:43:33  NOTI          Length: 144
17:43:33  NOTI          Flags: 0x40 (-P--)
17:43:33  NOTI          Command Code: 16777214
17:43:33  NOTI          ApplicationId: 16777215
17:43:33  NOTI          Hop-by-Hop Identifier: 0x2AC1D6DD
17:43:33  NOTI          End-to-End Identifier: 0x0BEF2407
17:43:33  NOTI             {internal data}: src:(nil)(0) rwb:0x0 rt:0
   cb:0x0,0x0(0x0) qry:0x7fbf82c42e00 asso:0 sess:0x7fbf82c432c0
17:43:33  NOTI          AVP: 'Session-Id'(263) l=47 f=-M val="
   peer2.localdomain;1445159102;2;app_test"
17:43:33  NOTI          AVP: 'Test-AVP'(16777215) vend='app_test
   vendor'(999999) l=16 f=V- val=438792350 (0x1a27709e)
17:43:33  NOTI          AVP: 'Origin-Host'(264) l=25 f=-M val="
   peer1.localdomain"
17:43:33  NOTI          AVP: 'Origin-Realm'(296) l=19 f=-M val="
   localdomain"
17:43:33  NOTI          AVP: 'Result-Code'(268) l=12 f=-M val='
   DIAMETER_SUCCESS' (2001 (0x7d1))
```

6. When done, press Ctrl-C in each terminal window to halt the peers.

This format is the most readable for humans, but it is also very space- (and therefore time-) consuming. Another option is to use the value 0x0040 in the configuration line, so that each message is written on a single line and may be reformatted post-mortem. That may be a better option for pilot runs before actual deployment of an application, for example. Not only are the sent and received messages displayed by this mechanism, but also different internal events, such as peers connections and message routing.

B.9.3 Measuring Processing Time: *dbg_msg_timings.fdx*

The *dbg_msg_timings.fdx* extension measures the processing time of the local stack for incoming messages and the network time between sent requests and received answers. This is useful for checking that a local application is fast enough and for determining if more application threads are required.

1. By default, the **CMake** option `BUILD_MSG_TIMINGS` is set to ON so you do not have to reconfigure the build.
2. Include the following line in the *freeDiameter-1.conf* file:

```
LoadExtension = "extensions/dbg_msg_timings.fdx";
```

3. Start the same test as in previous section. You will see the following:

```
8:16:25  NOTI   [TIMING] ANS 0.000404 sec ->'peer2.localdomain':
     'Test-Answer'16777215/16777214 f:-P-- src:'(nil)' len:144 {C
     :263/1:47,V:999999/C:16777215/1:16,C:264/1:25,C:296/1:19,C
     :268/1:12}
```

4. When done, press Ctrl-C in each terminal window to halt the peers.

B.9.4 Viewing Queue Statistics: *dbg_monitor.fdx*

The *dbg_monitor.fdx* extension displays some statistics of the framework every 30 seconds. To add it:

1. By default, the **CMake** option `BUILD_DBG_MONITOR` is set to ON so you do not have to reconfigure the build.
2. Include the following line in the *freeDiameter-1.conf* file:

```
LoadExtension = "extensions/dbg_monitor.fdx";
```

3. Start the same test as in previous section. You will see the following on peer1:

```
18:22:02  NOTI   [dbg_monitor] Dumping queues statistics
18:22:02  NOTI    Global 'Local delivery': cur:0/25, h:1, T:4 in
     0.000061s (65573.77items/s), blocked:0.000003s, last
     processing:0.000018s
18:22:02  NOTI    Global 'Total received': cur:0/20, h:1, T:4 in
     0.000060s (66666.67items/s), blocked:0.000003s, last
     processing:0.000011s
18:22:02  NOTI    Global 'Total sending': cur:0/30, h:1, T:4 in
     0.000113s (35398.23items/s), blocked:0.000004s, last
     processing:0.000015s
18:22:02  NOTI    {peer}(@0x7f86f9500580): peer2.localdomain [
     STATE_OPEN, cnt:0sr,0pa] rlm:localdomain ['freeDiameter'
     10201]
18:22:02  NOTI    'Events, incl. recept'@'peer2.localdomain': cur
     :0/0, h:1, T:6 in 0.000073s (82191.78items/s), blocked
     :0.000007s, last processing:0.000010s
18:22:02  NOTI    'Outgoing'@'peer2.localdomain': cur:0/5, h:1, T:4
     in 0.000058s (68965.52items/s), blocked:0.000003s, last
     processing:0.000011s
18:22:02  NOTI   [dbg_monitor] Dumping servers information
18:22:02  NOTI    {server}(@0x7f86f9500260)'{----} TCP srv
     [0.0.0.0]:3868 (4)': TCP, NotSecur(0), Thread running'pending
     connections'(@0x7f86f9500c40): items:0,0,5 threads:5,0 stats
     :0/0.000000,0.000000,0.000000 thresholds:0,0,0,0x0,0x0,0x0
{server}(@0x7f86f9440f20)'{----} TCP srv [0.0.0.0]:5658 (6)': TCP,
     Secur(1), Thread running'pending connections'(@0x7f86f9440f70)
     : items:0,0,5 threads:5,0 stats:0/0.000000,0.000000,0.000000
     thresholds:0,0,0,0x0,0x0,0x0
```

```
{server}(@0x7f86f96001a0)'{----} TCP srv [::]:3868 (7)': TCP,
    NotSecur(0), Thread running'pending connections'(
    @0x7f86f96001f0): items:0,0,5 threads:5,0 stats
    :0/0.000000,0.000000,0.000000 thresholds:0,0,0,0x0,0x0,0x0
{server}(@0x7f86f9701670)'{----} TCP srv [::]:5658 (8)': TCP, Secur
    (1), Thread running'pending connections'(@0x7f86f9701740):
    items:0,0,5 threads:5,0 stats:0/0.000000,0.000000,0.000000
    thresholds:0,0,0,0x0,0x0,0x0
```

4. When done, press Ctrl-C in each terminal window to halt the peers.

The first part of the output shows the states of the queues. This can help to identify the location of bottlenecks in a given deployment and adjust the number of consuming threads accordingly. It shows the global queues first and then each peer-specific queue.

The second part of the output shows the server sockets information. It displays the number of pending connections and may be used to detect some denial-of-service attacks, or identify some issues on very busy servers.

B.9.5   Understanding Routing Decisions: *dbg_rt.fdx*

The *dbg_rt.fdx* extension relates to the routing mechanism of `freeDiameter`. When a Diameter request message is to be sent, a specific thread selects the message's next hop. The algorithm is fairly complex. To avoid a lot of manual configuration, which is usually error prone, `freeDiameter` attempts to manage this task as reasonably and autonomously as possible. As a result, the default behavior of the stack takes into account a number of parameters such as the state of the connection, the applications advertised by the remote peer, the routing AVPs of the message, including `Destination-Host`, `Destination-Realm`, `Route-Record`, and the decorated Network Access Identifier (NAI) in `User-Name`.

Additional extensions can influence or change the routing decisions made by the framework; an example of such an extension is *rt_load_balance.fdx*, which can help to create a load balancer relay. All extensions with names starting with *rt_** are related to the routing mechanism.

To be as flexible as possible, the `freeDiameter` routing mechanism within the framework builds a list of the possible target peers, excluding peers that are temporarily disconnected, then allocates a score to each candidate based on different parameters, as described above. The list is then passed to each routing extension, which can freely change the score of different candidates. Once all routing extensions have been invoked, the framework orders the list and sends the message to the peer with the highest score.

The purpose of the *dbg_rt.fdx* extension is to help understand routing decisions. For each request being routed, it simply outputs the list of candidate peers and their score to the console. To enable:

1. By default, the **CMake** option `BUILD_DBG_RT` is set to `ON` so you do not have to reconfigure the build.
2. Include the following line in the *freeDiameter-1.conf* file:

```
LoadExtension = "extensions/dbg_rt.fdx";
```

Note that the simple test application that we have been interacting with has only one peer and thus does not provide a useful example of this extension. It is fairly easy to extend this simple setup with two additional peers in order to obtain a result similar to the following.

```
SEND 2ae8944a to 'localdomain' (-)
22:30:38   DBG   [dbg_rt] OUT routing message: 0x7fe2f951c1f0
22:30:38   DBG   'Test-Request'
 Version: 0x01
 Length: 20
 Flags: 0xC0 (RP--)
 Command Code: 16777214
 ApplicationId: 16777215
 Hop-by-Hop Identifier: 0x00000000
 End-to-End Identifier: 0x17EF04C4
 {internal data}: src:(nil)(0) rwb:0x0 rt:0 cb:0x107b1ca10,0x0(0
   x7fe2f951c420) qry:0x0 asso:0 sess:0x7fe2f951c080
 AVP: 'Session-Id'(263) l=8 f=-M val="peer1.localdomain;1481380222;1;
   app_test"
 AVP: 'Destination-Realm'(283) l=8 f=-M val="localdomain"
 AVP: 'Origin-Host'(264) l=8 f=-M val="peer1.localdomain"
 AVP: 'Origin-Realm'(296) l=8 f=-M val="localdomain"
 AVP: 'Test-AVP'(16777215) vend='app_test vendor'(999999) l=16 f=V-
   val=719885386 (0x2ae8944a)
22:30:38   DBG   [dbg_rt] Current list of candidates (0x7fe2f951c1f0):
   (score - id)
22:30:38   DBG   [dbg_rt]   13 -       peer2.localdomain
22:30:38   DBG   [dbg_rt]   13 -       peer3.localdomain
22:30:38   DBG   [dbg_rt]   13 -       peer4.localdomain
22:30:38   DBG   SENT to 'peer2.localdomain': 'Test-Request
   '16777215/16777214 f:RP-- src:'(nil)' len:152
   {C:263/1:47,C:283/1:19,C:264/1:25,C:296/1:19,V:999999/C
   :16777215/1:16}
```

B.9.6   The Interactive Python Shell Extension: *dbg_interactive.fdx*

Since one of the design goals of `freeDiameter` was to enable improvements to the Diameter protocol by allowing easy prototyping of new Diameter applications, the *dbg_interactive.fdx* extension was created to allow developers with a limited knowledge of the C language to write `freeDiameter` extensions using the Python language interactively. The conversion to and from Python adds some overhead and a factor of instability; for this reason it is not recommended to use this mechanism for deploying in real production environments, but for prototyping or research activities, using the interpreter is a nice option.

1. Ensure that you have the *swig* and *swig-python* libraries installed. SWIG is a tool that connects programs written in C and C++ with a variety of high-level programming languages like Python.

2. Enable `BUILD_DBG_INTERACTIVE` in your **CMake** configuration:

```
$ cmake -BUILD_DBG_INTERACTIVE:BOOL=ON ../src
```

3. Run **make**:

```
$ make
```

4. Include the following line in the *freeDiameter-1.conf* file:

```
LoadExtension = "extensions/dbg_interactive.fdx";
```

5. Start the stack:

```
$ freeDiameterd/freeDiameterd -c freeDiameter-1.conf
```

The interactive Python interpreter, which interacts directly with the framework, will start:

```
Starting interactive python interpreter [experimental].
16:56:00  Please use Ctrl-D to exit.
NOTI    Example syntax:
freeDiameterd daemon initialized.   >>> print cvar.fd_g_config.
   cnf_diamid
  'peer1.localdomain'

Python 2.7.10 (default, Jul 30 2016, 18:31:42)
[GCC 4.2.1 Compatible Apple LLVM 8.0.0 (clang-800.0.34)] on darwin
Type "help", "copyright", "credits" or "license" for more information.
>>>
```

The Python interface is described in *dbg_interactive.py.sample* found in the `src/-doc` directory, with examples for each service. This interface may not be complete, but it can be extended fairly easily.

6. Press Ctrl-D to stop the interpreter.

It becomes possible to create `freeDiameter` structures from Python and to register Python functions as callbacks to `freeDiameter` events. The following example shows how to create an AVP:

```
>>> gdict = cvar.fd_g_config.cnf_dict
>>> avp_model = gdict.search ( DICT_AVP, AVP_BY_NAME, "Origin-Host")
>>> avp_instance = avp(avp_model)
>>> val = avp_value()
>>> val.os = "my.origin.host"
>>> avp_instance.setval(val)
>>> avp_instance.dump()
'Origin-Host'(264) l=8 f=-M val="my.origin.host"
```

The first line retrieves a reference to the dictionary that is loaded in the framework. This dictionary contains the base protocol objects defined in the framework as well as any objects loaded by extensions such as *dict_sip.fdx*. The second line retrieves the definition of the `Origin-Host` AVP in this dictionary. The third line creates a new, blank instance of this AVP. The following three lines create and assign a value to the AVP. Finally, the `dump()` function shows the the content of the AVP as defined in the framework.

B.10   Further Reading

For more details on `freeDiameter`, and for more information on dependencies, visit http://www.freediameter.net/.

Reference

1 R. Hinden and S. Deering. IP Version 6 Addressing Architecture. RFC 4291, Internet Engineering Task Force, Feb. 2006.

Appendix C

The `freeDiameter` Framework

C.1 Introduction

The design of `freeDiameter` follows the principles of the two Diameter layers: the core implements the peer-to-peer messaging layer that manages the delivery of the Diameter messages and maintains connections between neighboring Diameter peers. `freeDiameter` has a higher layer running the Diameter applications that convey the service data and inherit the properties of the lower layer. `freeDiameter` allows applications to use the Diameter network to convey their messages.

When a `freeDiameter` node receives a Diameter message from a peer on a network interface, its security layer decrypts it and then the node sanity checks the message. If the message is a Diameter base protocol message, then it is handled locally in the context associated with the connection.

Other messages are sent to a global queue of incoming messages. An asynchronous task called *libfdcore* picks the messages from this queue and decides whether they should be forwarded to another peer, in which case they are stored in the queue of outgoing messages, or should be handled locally, and therefore they are stored in the queue of local messages.

Another asynchronous task picks the messages from the local queue and dispatches them to the appropriate application code that registered to receive these messages. The application can answer incoming requests and create new messages. When the application hands a message to the framework for sending, it is stored in the global queue of outgoing messages.

A third asynchronous task picks the messages from the outgoing message queue, decides to which Diameter peer they should be sent, and stores them in the appropriate queue of messages being sent to this peer. After sending, in case of requests, a copy is stored in the list of sent messages to be able to match incoming answers and to resend the messages if needed.

C.2 Framework Modules

The `freeDiameter` framework is organized into the following modules:

The *libfdproto* library
   This is a collection of stateless functions that allow the manipulation of Diameter objects: messages and AVPs, dictionary, sessions. Since this library is largely

Diameter: New Generation AAA Protocol – Design, Practice, and Applications, First Edition.
Hannes Tschofenig, Sébastien Decugis, Jean Mahoney and Jouni Korhonen.
© 2019 John Wiley & Sons Ltd. Published 2019 by John Wiley & Sons Ltd.

independent of the rest of the framework, it can be used in other software projects that need a simple Diameter message parser.

The *libfdcore* library

The core part of the framework is implemented in *libfdcore*. This library manages the connections to other Diameter peers, the routing of incoming and outgoing messages, and the loading of extensions described later in this section. It contains the complete implementation of the Diameter base protocol, the lower layer of Diameter as discussed in Chapter 3. This module can be easily embedded to add Diameter client capability to larger projects (e.g., virtual private network (VPN) gateways).

The **freeDiameterd** executable

When the framework is not embedded in another application, it can be used with the **freeDiameterd** executable. This package is a simple application that initializes *libfdcore* and waits for its termination. This is typically used on a Diameter server, where **freeDiameterd** can be installed as a system service.

C.3 `freeDiameter` API Overview

The complete source of documentation for `freeDiameter` API can be found in the comments of the header files in the `src/include/freeDiameter` directory:

libfdproto.h contains the API of the library with the same name; it is mostly related to manipulating data structures related to the Diameter protocol (messages, AVPs, etc.).

libfdcore.h describes the API to the main framework library, for starting or stopping the framework.

extension.h provides definitions that are useful when creating a `freeDiameter` extension, in particular the method to register the extension in the framework, including possible dependencies.

freeDiameter-host.h and *version.h* are two headers generated by **CMake**, and they contain the values of configurable options and system capabilities. These headers do not provide an API per se, but their contents may be useful when writing an extension.

You can find the code of a simple extension, with client and server sides, in the `extensions/test_app/` directory of the source code distribution. This can be used as a starting point for developing your own code and seeing how to use the `freeDiameter` API.

The main interactions of applications implementing a Diameter application with the `freeDiameter` framework are as follows:

- For a Diameter client, the application creates a Diameter message as a reaction to an application-specific stimulus, for example a new user connecting, and then uses the framework to send this message to a server. When received, the corresponding answer is automatically passed back to the application.
- For a Diameter server, it registers a function callback that will receive all incoming requests that belong to the specific Diameter application. When such request is received, the application handles the application-specific logic, then creates an answer and sends it back.
- Diameter agents are implemented similarly by registering functions to act upon specific events of `freeDiameter` and modify the default behavior.

C.3.1 *libfdproto.h*

This file defines the API related to a number of services provided by the library. The main services are DEBUG for the management of logging, DICTIONARY for a set of types and functions for managing a dictionary of Diameter definitions such as messages and AVPs, and MESSAGES for manipulating data structures related to Diameter messages. You should go through the file to learn what is available there before starting development, but we will list here a few of the most useful definitions for your extension code.

To add logs that can be turned on/off depending on the verbosity level, use the macro LOG(level, ...), which works like printf. For example:

```
LOG(FD_LOG_NOTICE, "my_ext starting with: %s", conffile);
```

When you call a function that may return an error under abnormal conditions, you may use the macro CHECK_FCT() or CHECK_SYS(), depending on the return convention, to take care of error checking and logging for you.

It is common for an extension to deal with binary strings and in particular DiameterIdentities, therefore a number of functions are provided as helpers, such as fd_os_almostcasesrch(), to compare and order binary strings with some considerations related to case sensitivity.

Diameter relies on AVP data structures, however the type of AVP is not explicit in the protocol, therefore an implementation must maintain a dictionary of objects in order to parse messages properly, for example to distinguish between an OctetString value or a grouped AVP. freeDiameter extends the semantics of the dictionary to automate many tasks such as validating a message against ABNF, parsing semantics of an AVP value, etc. You will need to use the functions related to dictionary in any extension, such as fd_dict_search(), which is used to get a reference to a dictionary object, and use it afterwards during the parsing or creation of messages. fd_dict_new() is used to add new definitions to the dictionary, typically used in *dict_*.fdx* extensions by convention, but nothing prevents another extension from using it as well. The header also contains some definitions of commonly used constant values, such as ER_DIAMETER_SUCCESS, which is useful when parsing an Error-Code AVP value.

```
/* Retrieve a reference on Result-Code AVP in dictionary */
struct dict_object * d_res_code = NULL;
CHECK_SYS(
        fd_dict_search(fd_g_config->cnf_dict,
                DICT_AVP,
                AVP_BY_NAME,
                "Result-Code",
                &d_res_code,
                ENOENT )
);

/* Create a Test-Payload AVP in dictionary */
struct dict_object * d_test_payload = NULL;
struct dict_avp_data data;

data.avp_code = 0x1234;
data.avp_vendor = 0x9988;
data.avp_name = "Test-Payload";
data.avp_flag_mask = AVP_FLAG_VENDOR;
```

```
data.avp_flag_val = AVP_FLAG_VENDOR;
data.avp_basetype = AVP_TYPE_OCTETSTRING;

CHECK_FCT(
        fd_dict_new(fd_g_config->cnf_dict,
                    DICT_AVP,
                    &data,
                    NULL,
                    &d_test_payload )
);
```

`freeDiameter` provides support functions for handling Diameter sessions. An extension can register data with a session and retrieve it later easily when a new message with the same `Session-Id` AVP value is handled, using `fd_sess_state_retrieve()`. This mechanism enables different extensions to manage their own data with each session, without collisions. However, the state machine related to user sessions is application-specific and needs to be implemented in the extension itself.

One of the most used features is the set of functions related to messages handling, such as `fd_msg_avp_new()` to create a new AVP instance, `fd_msg_browse()` to navigate within a Diameter message instance, and `fd_msg_dump_full()` to log the content of a message in human-readable format. Another useful function is the `fd_msg_answ_getq()` that retrieves a reference to the request message corresponding to an answer that has been received. The value of AVP instances is set with `fd_msg_avp_setvalue()` in general or `fd_msg_avp_value_encode()` when the dictionary specifies a method to convert an AVP value to a specific format. An example of this mechanism is used with the Diameter type address that can directly be mapped to and from a `sockaddr` structure.

```
/* Create an answer to a message 'msg' */
CHECK_FCT(
        fd_msg_new_answer_from_req(fd_g_config->cnf_dict, &msg, 0) );

/* Create a Test-Payload AVP instance */
struct avp * a;
union avp_value val;

val.os.data = "123456789";
val.os.len = 9;

CHECK_FCT( fd_msg_avp_new ( d_test_payload, 0, &a ) );
CHECK_FCT( fd_msg_avp_setvalue( a, &val ) );

/* Add this AVP in the message 'msg' as last AVP */
CHECK_FCT( fd_msg_avp_add( msg, MSG_BRW_LAST_CHILD, a ) );
```

Extensions almost always need to register to receive specific Diameter messages. This is done with the `fd_disp_register()` function. Note that this does not impact the list of applications advertised in the Capability-Exchange handshake by the local peer. To add an application to the list of advertised applications, refer to the *libfdcore* API.

```
/* Register the function my_cb to handle incoming my_req messages */
struct disp_when data;
struct disp_hdl * my_tr;
```

```
data.app = d_my_app;        /* dict_object */
data.command = d_my_cmd_r; /* dict_object */

CHECK_FCT(
        fd_disp_register( my_cb, DISP_HOW_CC, &data, NULL, &my_tr) );
```

C.3.2 *libfdcore.h*

There are two ways to use the *libfdcore* library: embed it in a larger application or load it from the **freeDiameterd** application. In either case, an application that wants to send or receive Diameter messages will need to call the functions of the framework. This can be done directly by the application embedding *libfdcore*. This is a common approach when adding Diameter support to a single application.

A number of functions, such as fd_core_start(), are used to control the framework and are not called from extensions but from the application embedding freeDiameter. However, the fd_core_waitstartcomplete() function may be useful to extensions that are starting a thread during their initialization, such as the *dbg_interactive.fdx* discussed earlier in this chapter.

The instance configuration is available directly through the global variable fd_g_config. It contains information about the dictionary (cnf_dict) or the main event queue (cnf_main_ev). Extensions can read all the configuration information but should never need to change this data directly.

One of the main roles of the *libfdcore* library is the management of the connections to other Diameter entities in the network. Each remote peer is represented in the local instance by a peer structure, which contains a peer_info member with information readable by extensions, such as the peer's DiameterIdentity supported applications, etc. While peers can be declared in the configuration file, they can also be dynamically added by extensions via the fd_peer_add() call. An extension may also need to list all the peers represented in the local framework, which can be achieved by enumerating the entries of fd_g_peers as follows:

```
/* Take a read lock on the list */
CHECK_FCT( pthread_rwlock_rdlock(&fd_g_peers_rw) );

/* For each entry */
for ( li = fd_g_peers.next; li != &fd_g_peers; li = li->next) {
        struct peer_hdr * p = (struct peer_hdr *)li->o;

/* Read what you need from p here.
        * You can use fd_peer_get_state() for example
        * to discover if the peer is in OPEN state and
        * messages can be sent to it. */
}

/* Release the lock */
CHECK_FCT( pthread_rwlock_unlock(&fd_g_peers_rw) );
```

Finally, while the default behavior of freeDiameter is to reject incoming connections from unknown peers, an extension may register a callback with fd_peer_validate_register() to accept incoming connections selectively. See the *acl_wl.fdx* extension for reference, for example.

In addition to managing connections, this library also handles messages to and from these peers. An extension must use `fd_msg_send()` or `fd_msg_send_timeout()` to send a message that it has created. Note that these functions do not take a destination peer explicitly; the framework will route the message implicitly according to the rules that have been described earlier. When the message being sent is a request, the extension can pass a callback that will be called when a corresponding answer is received. The library also provides a few high-level functions to manipulate messages easily, such as `fd_msg_rescode_set()`, which adds results information to a message, the `Result-Code` AVP, and, if applicable, the "E" flag, `Error-Message`, `Error-Reporting-Host`, and `Failed-AVP` AVPs.

As mentioned previously, an extension needs to register a callback to receive specific messages; in addition it needs to explicitly register the applications that need to be advertised during the capability-exchange handshake. This is done with the `fd_disp_app_support()` function.

```
CHECK_FCT(
        fd_disp_app_support(d_my_appli,
                            d_my_vendor,
                            1, /* advertise as authentication app */
                            0) /* don't advertise as accounting */
);
```

To change the routing behavior of `freeDiameter`, an extension can use either `fd_rt_fwd_register()` or `fd_rt_out_register()` to register the callback to be called during processing of forwarded messages (typically to implement a Diameter Proxy function) or outgoing messages (to implement a load balancer).

An extension can register for external triggers through the framework using the function `fd_event_trig_regcb()`. This is how the *test_app.fdx* extension registers to trigger a message when the daemon receives SIGUSR1.

`freeDiameter` implements a number of hooks in the framework that enable an extension to register callbacks on specific conditions. Hooks can be used to implement debugging functionality (e.g., the *dbg_msg_dumps.fdx* extension) or monitoring features. It can also be useful to implement advanced processing of error messages in some situations, or to keep track of the connection status to different peers. The extension uses `fd_hook_register()` to register a callback on the events they are interested in:

```
struct fd_hook_hdl *hdl;

CHECK_FCT( fd_hook_register(HOOK_MASK( HOOK_MESSAGE_RECEIVED,
                                       HOOK_MESSAGE_SENT,
                                       HOOK_DATA_RECEIVED
                                       ),
                            my_hook_cb,
                            NULL,
                            NULL,
                            &hdl ) );
```

The framework maintains statistics on its internal message queues, and these can be consulted with `fd_stat_getstats()`. It is useful for monitoring, such as with the *dbg_monitor.fdx* extension, or as a basis for implementing Simple Network Management Protocol (SNMP) [1] support. In addition, `fd_stat_getstats()` could be used to implement automatic responses to a peer becoming overloaded.

C.3.3 *extension.h*

This header defines the macro EXTENSION_ENTRY() to simplify the process of defining the entry point for an extension. If you are writing an extension *my_ext.fdx* that uses definitions from *dict_eap.fdx*, and your starting point is the function:

```
static int my_init(char * conffile);
```

then you would add the following in your main C file:

```
EXTENSION_ENTRY("my_ext", my_init, "dict_eap");
```

This ensures that the framework will be able to call your entry function when this extension is listed in the main configuration file.

  If you need to clean some state when the framework is stopping, you can also define a function with this exact name and prototype, and it will be called automatically:

```
void fd_ext_fini(void);
```

C.4   freeDiameter Architectures

There are two possible architectures when working with freeDiameter. One approach is to use freeDiameter extensions. These extensions are plug-ins that *libfdcore* loads during its initialization, based on its configuration. The number of extensions that can be loaded is unlimited. It is common practice to split a feature into several small extensions that can be easily reused in different contexts. Such small extensions include, for example, those that only add the definitions of AVPs and commands for a specific Diameter application in the freeDiameter dictionary. These dictionary extensions can be reused in each Diameter node – client, proxy, or server – that deals with this Diameter application. These extensions improve flexibility and reusability; however, they are also freeDiameter-centric. This architecture is typically the one to choose for Diameter servers and proxies, whose main functionality is the Diameter processing itself.

  The alternative architecture is to embed freeDiameter into an application. This is generally done on the Diameter client side. From the Diameter perspective, there is only one application to be supported and the routing is usually simple, while from the application perspective, the main logic is not related to Diameter but to the service itself. For example, an 802.1X authenticator application focuses on managing the ports and the state machines of the supplicants, with options to authenticate the supplicants either locally or via an AAA mechanism, such as RADIUS or Diameter, therefore in such applications it is desirable that freeDiameter is just an optional module.

  Such considerations were at the core of the design of freeDiameter. The **freeDiameterd** application is an example of such an application embedding freeDiameter, with little logic beyond what is needed to initiate the library. Once an application embeds the freeDiameter libraries, it can implement its Diameter messaging either as an extension loaded by the framework or by calling the framework API from the main application. There are two points that require some care. The parsing of the configuration file is not flexible at the time of writing this text, which means your application will need a dedicated configuration file for Diameter subsystem. It is on freeDiameter roadmap to change this, but with a low priority. The other item that requires care is the timing at which you may call different functions from `freeDiameter`. For example, it is not possible to send a message before the framework is fully initialized and has established a connection to at least one of the backend servers. There are some helper functions in `freeDiameter` that provide feedback on these events to the application so that it can adapt its behavior accordingly.

Reference

1 D. Harrington, R. Presuhn, and B. Wijnen. An Architecture for Describing Simple Network Management Protocol (SNMP) Management Frameworks. RFC 3411, Internet Engineering Task Force, Dec. 2002.

Glossary

access point name A gateway between a mobile network and another computer network, such as the Internet. A mobile device making a data connection must be configured with an APN so that the mobile network can determine which IP address to assign and which security methods should be used. An APN identifies the network that the user wants to communicate with, and may also include the service requested by the user

accounting The tracking of the user's consumption of resources for billing, auditing, and/or system planning

application error An error in Diameter that occurs due to an issue with an application-specified function, like a user authentication error

application programming interface A clearly defined communication interface between two software components

authentication The verification that a user who is requesting services is a valid user of the network services requested

authorization The determination of whether requested services can be granted to a user who has presented an identity and credentials based on their authentication, service request, and system state

Diameter node A software application that implements the Diameter protocol and is a participant in a Diameter session

Diameter peers Two Diameter nodes that have a direct transport protocol connection between them

DiameterIdentity An AVP data format for capturing either a realm, which helps determine whether a message can be handled locally or needs to be routed, or fully qualified domain name, which identifies a Diameter node. DiameterIdentity is the data format for the following common AVPs: `Destination-Host`, `Destination-Realm`, `Origin-Host`, `Origin-Realm`

downstream This term identifies the direction a Diameter message travels from the Diameter client towards the server that will handle the request

finite state machine A model of computer program behavior that can be in exactly one of a finite number of states at any given time. A specific event or trigger causes the state to transition to a new state

head-of-line blocking A situation where transport protocol level retransmissions delay the delivery of all subsequent messages until the retransmitted message gets through

Diameter: New Generation AAA Protocol – Design, Practice, and Applications, First Edition.
Hannes Tschofenig, Sébastien Decugis, Jean Mahoney and Jouni Korhonen.
© 2019 John Wiley & Sons Ltd. Published 2019 by John Wiley & Sons Ltd.

home The administrative domain with which the user maintains an account relationship

Home Subscriber Server A 3GPP network element that stores identity, authentication, and policy information about mobile subscribers

Hop-by-Hop Identifiers Used between adjacent Diameter nodes to map answer messages to request messages

IMSI international mobile subscriber identity, a concatenation of mobile network code (MNC), mobile country code (MCC) values, and mobile subscriber identification number (MSIN) that is unique to each subscriber in a mobile network

Internet Engineering Task Force An international community of network designers, operators, vendors, and researchers that develop open, voluntary Internet standards

Mobility Management Entity A 3GPP network element that provides authentication, network security, and roaming capability to mobile devices

network access server A generic term for the end user's entry point to a network. A NAS provides services on a per-user basis, based on authentication, and ensures the service provided is accounted for

network byte order The order in which bytes in a multi-byte number are transmitted with the most significant byte first. Also known as "Big Endian". Fields in a Diameter message are transmitted in this order

peer control block A memory control structure that contains the transport connection state of a specific peer and is part of the peer state machine

peer state machine A finite state machine that must be observed by all Diameter implementations and that keeps track of a peer's connection status

peer table A data structure internal to a Diameter node that holds the information of adjacent remote Diameter peers with which the Diameter node has transport-level connections established

public key infrastructure A public key infrastructure (PKI) is a set of roles, policies, and procedures needed to create, manage, distribute, use, store, and revoke digital certificates and manage public-key encryption

public land mobile network A network that is established and operated to provide mobile telecommunications services to the public

realm An administrative domain

remote access server A combination of hardware and software that allows access by authorized users to a network

Request for Comments A document created by the IETF that has gone through the IETF's review and approval process. RFCs cover not only Internet standards, but also research discussions and best current practices

session state State that allows a Diameter node to track all authorized, active sessions

Single Radio Voice Call Continuity A 3GPP solution that provides voice call continuity for calls that are anchored in IMS when the UE is capable of transmitting/receiving on only one of either the packet-switched or circuit-switched access networks at a given time

subscriber identity module A chip that securely stores a mobile phone's International Mobile Subscriber Identity (IMSI) and authentication key

transaction state The state maintained by a Diameter node in order to match answers with requests

upstream The direction a Diameter message travels from the answering Diameter server towards the Diameter client

VoLTE Voice over Long-Term Evolution, a standard for high-speed communication over the wireless network that no longer depends on the circuit-switched network

Index

Diameter: New Generation AAA Protocol – Design, Practice, and Applications, First Edition.
Hannes Tschofenig, Sébastien Decugis, Jean Mahoney and Jouni Korhonen.
© 2019 John Wiley & Sons Ltd. Published 2019 by John Wiley & Sons Ltd.